Innovations in
Titanium
Technology

INNOVATIONS IN TITANIUM TECHNOLOGY

TMS Member Price: $67   TMS Student Member Price: $54

List Price: $97

Related Titles
- *Cost-Affordable Titanium,*
  edited by F.H. (Sam) Froes, M. Ashraf Imam and Derek Fray
- *TMS Landmark Paper Series: Titanium Metal Matrix Composites,*
  edited by Warren H. Hunt Jr.

HOW TO ORDER PUBLICATIONS

For a complete listing of TMS publications, contact TMS at
(800) 759-4TMS or visit the TMS Document Center at http://doc.tms.org:

- Purchase publications conveniently online.
- View complete descriptions, tables of contents and sample pages.
- Find award-winning landmark papers and reissued out-of-print titles.
- Compile customized publications that meet your unique needs.

MEMBER DISCOUNTS

TMS members receive a 30% discount on TMS publications. In addition, members receive a free subscription to the monthly technical journal *JOM* (both in print and online), discounts on meeting registrations, and additional online resources to name a few of the benefits. To begin saving immediately on TMS publications, complete a membership application when placing your order in the TMS Document Center at http://doc.tms.org or contact TMS.

Telephone: (724) 776-9000 / (800) 759-4TMS
E-mail: membership@tms.org or publications@tms.org
Web: www.tms.org

Innovations in Titanium Technology

Proceedings of a symposia sponsored by
the Materials Processing & Manufacturing Division of
TMS (The Minerals, Metals & Materials Society)

TMS 2007 Annual Meeting & Exhibition
Orlando, Florida, USA
February 25-March 1, 2007

Edited by

Mehmet N. Gungor

M. Ashraf Imam

F.H. (Sam) Froes

A Publication of

A Publication of **The Minerals, Metals & Materials Society (TMS)**
184 Thorn Hill Road
Warrendale, Pennsylvania 15086-7528
(724) 776-9000

Visit the TMS Web site at
http://www.tms.org

Library of Congress Catalog Number 2007921104
ISBN Number 978-0-87339-665-3

If you are interested in purchasing a copy of this book, or if you would like to receive the latest TMS publications catalog, please telephone (724) 776-9000, ext. 270, or (800) 759-4TMS.

INNOVATIONS IN TITANIUM TECHNOLOGY SYMPOSIUM TABLE OF CONTENTS

Innovations in Titanium Technology Symposium

Low Cost Materials and Processing

Novel Materials and Processes I

Novel Materials and Processes II

Advances in Materials Processing

Advances in Alloy Development

Microstructure and Properties I

Microstructure and Properties II

PREFACE

This book constitutes the proceedings of the symposium "Innovations in Titanium Technology" held at the TMS 2007 Annual Meeting & Exhibition in Orlando, Florida, United States. This symposium was a follow-up to "Cost Affordable Titanium" held in Charlotte, North Carolina, United States, in 2004. The goal of the 2007 symposium was to report on innovations in titanium and its alloys science and technology with emphasis on cost reductions. The diverse mix of papers presented in these proceedings reflects the breadth of activities in titanium research and development. The opening overview paper, by the editors, presents a broad view of present and projected titanium cost reductions, both in titanium production and in subsequent processing. What is particularly clear is that the impetus for broadening markets for titanium is very dependent on cost reductions, particularly in applications such as the cost sensitive automobile industry. The symposium was organized into seven sessions which are likewise represented in this publication: Low Cost Materials and Processing, Novel Materials and Processes I & II, Advances in Materials Processing, Advances in Alloy Development, and Microstructures and Properties I & II.

The symposium organizers, who are also editors of this volume, would like to thank everyone who helped to make the event a success, particularly the session chairs: Dr. James Sears, South Dakota School of Mines and Technology; Dr. Patrick Martin, Air Force Research Laboratory MLLMD; Dr. Stephen Fox, TIMET; Taras Lyssenko, International Titanium Powder; Dr. Richard Dashwood, Imperial College London; Dr. Daniel Eylon, University of Dayton; Dr. Stephen Gerdemann, U.S. Department of Energy; Dr. Kuang-Oscar Yu, RMI Titanium; Dr. Charles Yolton, Crucible Materials Corporation; Dr. Ibrahim Ucok, Concurrent Technologies Corporation; and Dr. Catherine Wong, Naval Surface Warfare Center. The organizers would also like to thank the presenters, the authors of the papers contained in these proceedings, and the TMS Titanium Committee for sponsoring the symposium. Finally, the editors would like to recognize the great help received from Christina S. Raabe, Cheryl Moore and Marla K. Boots of TMS.

Mehmet N. Gungor, Sc.D., *Concurrent Technologies Corp.*

M. Ashraf Imam, D.Sc., *Naval Research Laboratory*

F.H. (Sam) Froes, Ph.D., *University of Idaho*

FOREWORD

This book constitutes the proceedings of the TMS symposium devoted to "Innovations in Titanium Technology" held at the 2007 TMS Annual Meeting in Orlando, Florida, U.S.A. This symposium is a follow-up to the symposium on "Cost Affordable Titanium" organized in Charlotte, North Carolina, U.S.A. in spring 2004. The goal of the present symposium was to report on innovations in titanium and its alloys science and technology with emphasis on cost reductions. The diverse mix of papers presented in these proceedings reflects the breadth of activities in the titanium R&D arena. The opening overview paper, by the editors, attempts to present a broad view of present and projected titanium cost reductions both in titanium production and in subsequent processing. What is particularly clear is that the impetus for broadening markets for titanium is very dependent on cost reductions, particularly in applications such as the cost sensitive automobile industry. The seven session symposium was organized into sessions on Low Cost Materials & Processing, Novel Materials & Processes I & II, Advances in Materials Processing, Advances in Alloy Development, and Microstructures and Properties I & II. The sections of this publication are separated in the same way.

The organizers of this symposium, who are also the editors of this volume, would like to thank everyone who helped to make the event a success, and particularly the session chairmen; Dr. James Sears – South Dakota School of Mines and Technology, Dr. Patrick Martin – Air Force Research Laboratory MLLMD, Dr. Stephen Fox – TIMET, Taras Lyssenko – International Titanium Powder, Dr. Richard Dashwood – Imperial College London, Dr. Daniel Eylon – University of Dayton, Dr. Stephen Gerdemann – U.S. Department of Energy, Dr. Kuang-Oscar Yu – RMI Titanium, Dr. Charles Yolton – Crucible Materials Corporation, Dr. Ibrahim Ucok – Concurrent Technologies Corporation and Dr. Catherine Wong-Naval Surface Warfare Center, the presenters and the authors of the papers contained in these proceedings. We would like to thank the TMS Titanium Committee for sponsoring this symposium. We would also like to recognize the great help we received from Christina S. Raabe, Cheryl Moore and Marla K. Boots of TMS.

Mehmet N. Gungor, ScD   M. Ashraf Imam, ScD   F.H. (Sam) Froes, PhD
Concurrent Technologies Corp. Naval Research Laboratory University of Idaho

ABOUT THE EDITORS

Mehmet N. Gungor is with Concurrent Technologies Corporation after working with the Science and Technology Center of Westinghouse Electric Corporation and PCC Composites. His work focused on advanced materials and processing research for advanced applications. Dr. Gungor is a member of The Minerals, Metals & Materials Society (TMS) and ASM International, and serves on several TMS committees including titanium, solidification and composites. He also served as a United Nations Advanced Materials Expert in Cordoba, Argentina. The editor of four books, Dr. Gungor also has more than 75 publications and two patents to his credit. He earned his Sc.D. in materials engineering from Massachusetts Institute of Technology.

M. Ashraf Imam is a member of the senior research staff in the Materials Science and Technology Division at the Naval Research Laboratory (NRL), Washington, D.C. He serves as a team leader in pursuing basic research on material structure-property relationship with emphasis on titanium alloy development and kinetics of transformation in solid phase. In addition to his work at NRL, Dr. Imam also holds the position of adjunct professor at The George Washington University in Washington, D.C., where he teaches materials-related subjects in the department of mechanical and aerospace engineering. He is a Fellow and member of the American Society for Metals International and also holds membership in both The Minerals, Metals & Materials Society and Sigma Xi. Dr. Imam received the 2003 George Kimbell Burgess Award from the Washington D.C. Chapter of the American Society for Metals. He earned his doctorate of science from The George Washington University.

F.H. (Sam) Froes is director of the Institute for Materials and Advanced Processes and head of the materials science and engineering department at the University of Idaho. He has more than 800 publications and 60 patents to his credit and has edited almost 30 books. Dr. Froes is a member of the All-Russian Academy of Natural Sciences and a Fellow of ASM International. Among his other accomplishments, he has received an award for service to powder metallurgy from the MPIF/APMI. Earlier in his career, Dr. Froes worked for the Air Force Materials Laboratory in Dayton, Ohio, where he progressed to special assistant to the Metals and Ceramics Division chief. He joined the lab after managing the titanium group at the Crucible Research Center, in Pittsburgh, Pennsylvania. Dr. Froes earned his doctorate in physical metallurgy from Sheffield University, United Kingdom.

Innovations in
Titanium
Technology

Low Cost Materials and Processing

COST AFFORDABLE TITANIUM – AN UPDATE

F.H. (Sam) Froes[1], Mehmet N. Gungor[2] and M.Ashraf Imam[3]

[1] Institute for Materials and Advanced Processes, University of Idaho; McClure Hall Room 437; Moscow, ID 83844-3026
[2] Concurrent Technologies Corporation, 100 CTC Drive, Johnstown, PA 15904
[3] Naval Research Laboratory; 4555 Overlook Avenue S.W.; Washington, DC 20375-5343

Keywords: Cost-affordable, extraction, fabrication, near net shapes

Abstract

Titanium is the "wonder" metal, which makes sense as the material of choice for a wide variety of applications. However because of its relatively high price – a result of extraction and processing costs- it is used basically only when it is the only choice; with the caveat that titanium has a bright "image" which can lead to use even when the economics are unfavorable. This paper will overview the potential areas which are amenable to cost reduction for titanium products. This will emphasize all steps in component fabrication from extraction and processing to fabrication of final parts.

Cost Reduction

The major thrust in the area of titanium development has been aimed at achieving cost reduction rather than developing alloys with enhanced properties (1). The cost of titanium compared to steel and aluminum is shown in Table I. Broadly speaking cost reduction can come from either reduction in the cost of production of the metal itself or from creative techniques for the fabrication of final components. Over the past few years there has been a lot of activity in the area of reduced cost titanium extraction processes. This is particularly a result of new and developing applications such as armor and auto use where a cost reduction could significantly increase use. However it must be kept in mind that in the big picture the cost of extraction is a small fraction of the total cost of a component fabricated by the cast and wrought (ingot metallurgy) approach, Figure 1.

Table 1: Cost of Titanium – A Comparison*

ITEM	MATERIAL ($ PER POUND)		
	STEEL	ALUMINUM	TITANIUM
ORE	0.02	0.10	0.22 (RUTILE)
METAL	0.10	1.10	5.44
INGOT	0.15	1.15	9.07
SHEET	0.30-0.60	1.00-5.00	15.00-50.00

*Contract prices. The high cost of titanium compared to aluminum and steel is a result of (a) high extraction costs and (b) high processing costs. The latter relating to the relatively low processing temperatures used for titanium and the conditioning (surface regions contaminated at

the processing temperatures, and surface cracks, both of which must be removed) required prior to further fabrication.

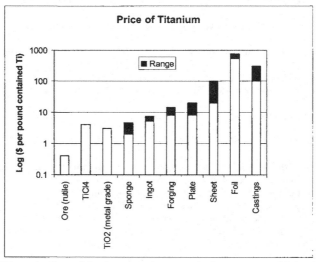

Figure 1. Cost of titanium at various stages of a component fabrication.

The possible market penetration resulting from cost reduction is demonstrated in a study of the potential applications for titanium (commercially pure) metal powder conducted by Norgate and Wellwood (2). This work assumed that there could be a 50% reduction in the cost of titanium CP (from a figure of $13.60 / lb) and developed the data shown in Table II for various applications. This represents a substantial greater than 1100% increase in titanium use – which seems unrealistically high. However it does demonstrate the great influence that the cost of titanium can have on the magnitude of applications.

Table II: Potential World-Wide Market For Commercially Pure Titanium.* (2)

APPLICATION	VOLUME (TONS/YEAR)
COOKWARE	39,000
MEDICAL IMPLANTS	1,000
ARCHITECTURE, BUILDING	
AND CONSTRUCTION	343,000
AUTOMOTIVE EXHAUST SYSTEMS	48,000
TUBING	290,000

*Assuming a cost reduction in titanium extraction and fabrication of 50%; emphasis on the powder metallurgy approach.

Examples of consumer goods fabricated from titanium are shown in Figure 2.

Figure 2. Titanium hammer head and knife handle (Courtesy Stiletto Tool and Buck Knives, respectively).

Extraction

In a review at the International Titanium Association (ITA) Conference in Scottsdale, AZ (Oct. 2005) of the various mainly extraction processes under development in various parts of the world, Dr. Ed Kraft noted that there are more than 20 new processes under development many from the oxide as a precursor [3]. This makes sense as the cost of the contained titanium is less in the oxide than, for example, in the intermediate compound (in Kroll/Hunter reduction) the tetrachloride, Table III. He noted that the costs of providing adequate purity and morphology of oxides must also be considered, as should the potential of halide recycling. Dr. Kraft also pointed out that a number of the new processes could result in products which eliminate some of the current ingot metallurgy processing steps (Figure 1), hence giving a greater cost reduction than from lower extraction costs per se. Examples are production of sheet and "chunky" shapes directly from powder.

Table III. Cost of Titanium Precursors

PRECURSOR	COST ($/lb)	COST OF CONTAINED Ti ($/lb)
TiO_2*	1.75	2.94
$TiCl_4$	1.00	4.00
Ti Sponge	5.44	5.44

*Metal grade

The approaches mentioned by Dr. Kraft are shown in Table IV, and include four processes being funded by DARPA, among these processes the FFC approach has been quoted as capable of producing sponge for as little as $1 per pound [4] compared to Kroll sponge at $3.50 per pound, however other analysis of this method [5] suggests a cost similar to the Kroll approach. Timet has experienced difficulties in scale-up of this process, but more promising results have been obtained by Norsk Ti.

The extraction process closest to commercialization is the Armstrong/ITP process (figure 3) which is scheduled for scale-up to 4 million lbs per year, late 2007. Perhaps the process with the greatest potential is the MER process, projected to produce Ti for a cost of $1.30/lb. Whether any of these processes will mature to the point where it supercedes the Kroll/Hunter approach remains to be seen.

Table IV. Titanium Extraction Processes (3)

TECHNIQUES	COMMENTS
FFC*	Oxide, electrolytic molten $CaCl_2$
MER*	Oxide, electrolytic
SRI*	Fluidized bed H_2 reduction of $TiCl_4$
BHP (Billiton, Aust.)	Oxide electrolytic, pre-pilot plant
Idaho Ti	Plasma quench, chloride
Ginatta, Italy	Electrolytic, chloride
OS (Ono, Japan)	Electrolytic/calciothermic oxide
MIR, Germany	Iodide reduction
CSIR, South Africa	Electrolysis of oxide
Okabe-I, Tokyo, Japan	Oxide, reduction by Ca
Okabe-II, Tokyo, Japan	Oxide, Ca vapor reduction
Vartech, Idaho	Oxide, Ca vapor reduction
Northwest Inst. For Non-Ferrous Metals	Innovative hydride-dehydride
CSIRO, Australia	Chloride, fluidized bed, Na.
Armstrong/ITP*	Chloride, continuous reduction with Na
DMR	Aluminothermic rutile feedstock
MIT	Oxide, electrolysis
QIT/Rio Tinto	Slag, electrolysis
Tresis	Argon plasma, chloride
Dynamet Technology	Low cost feedstock

* DARPA funded.

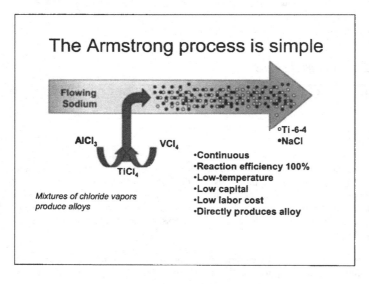

6

Figure 3. Schematic of the Armstrong/International Titanium Powder (ITP) Process indicating the production of Ti-6Al-4V alloy powder.

Primary Fabrication

In the melting arena, plasma arc melting (PAM) and electron beam melting (EBM) are cold hearth melting processes which provide the advantage of flexibility using various forms of low cost input materials, leading to reduced costs [6,7]. Specifically, PAM and EBM offer the potential to make single melt near net shapes (NNS) slabs and ingots and currently the U.S. Army, Navy and Air Force all have active programs in this area. Figures 4 and 5 show a schematic of the PAM slab casting process and an as-cast slab [8].

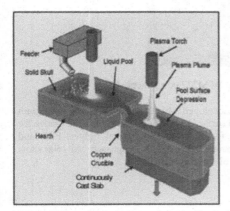 

Figure 4. Schematic of slab casting by melting process [8].

Figure 5. A 13" x 34" x 50" slab cast in plasma arc the furnace shown in Figure 4 [8].

These techniques also increase the time for which metal is molten hence increasing the chance of homogeneity and removal of defects such as the type I species (oxygen-nitrogen stabilized). This latter effect is particularly important for rotating quality stock produced from alloys such as Ti-6Al-4V, Ti-6Al-2Sn-4Zr-2Mo-Si and Ti-17.

Near Net Shape Processes

Castings

Castings are a cost-effective method of producing complex near net shapes (NNS). Castings offer cost reduction by minimizing machining, reducing part count and avoiding part distortion from fabrications such as machining and welding. Traditionally the investment casting technique has been used for aerospace (mainly gas turbine engine) parts, and rammed graphite castings for large chemical processing industry components. One recent excellent demonstration of titanium castings use is in M777 Lightweight Howitzer applications where fabricated titanium parts have been converted into unitized investment castings to reduce cost and distortions (9-12). One good example of a cast component developed and implemented is shown in Figure 6 (12). This part reduced cost because 60 fabricated parts integrated into a single part (12).

Figure 6. Invesment cast Ti-6Al-4V M777 component (12).

Recently metal mold titanium castings have been developed which exhibit low cost, finer grain sizes and much lower alpha case (oxygen enriched surface regions). Already, variable blades for the F119 gas turbine engine (the power system for the advanced tactical fighter, F22) and automobile valves for Formula I racing cars have been made using this approach. At 1500 shots per die the process is well beyond the break-even point to be profitable.

Powder Metallurgy

Powder metallurgy (P/M) approaches to production of near net shapes (NNS) have been explored in various parts of the world, both blended elemental (BE) and prealloyed (PA) methods (PA) [4,13], as techniques for production of cost effective shapes (Figure 7). High integrity P/M components have been produced via PA in the USA from both conventional alloys and the intermetallic TiAl; however with virtually zero commercial applications. Creative advances in the use of BE components in Japan have included the Toyota Altezza valves [14]. In the USA hydrogenated Ti-6Al-4V leads to higher and more consistent densities; and at lower cost by innovative direct production of hydrogenated powder [4,13]. This approach is also amenable to incorporation of reinforcing particles into the titanium matrix (Figure 8) [15].

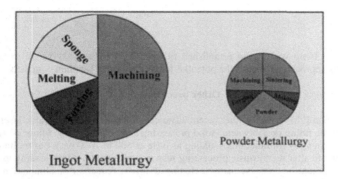

Figure 7. A comparison of the price of conventional ingot metallurgy product ("chunky" forging) with a powder metallurgy near net shape, not to scale.

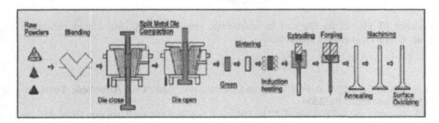

Figure 8. Processing steps for the blended elemental (BE) Ti-MMC engine valve production [15].

Another P/M NNS shape approach which is receiving increased attention is metal injection molding (MIM) – which is discussed in a paper in these proceedings (16). This process, developed from plastic injection molding, is capable of producing very complex parts generally less than 500 grams in weight in large numbers (Figures 9 & 10).

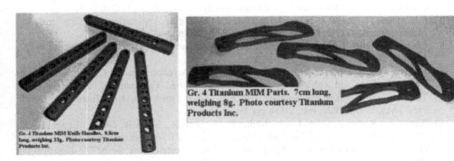

Figure 9. Titanium MIM Knife   Figure 10. Titanium MIM Parts.

9

Handles.

Thus while there is no widespread established titanium P/M industry as yet in the USA (or worldwide), this approach does offer the potential for cost-effective titanium compacts.

Other processes

Beyond casting and P/M products other recent advances in the NNS arena include laser forming and further into the future possibly semi-solid processing and cold spraying. Minor perturbations to existing titanium alloy formulations – adding as little as 500 to 1000 parts per million boron – have been shown to alter the intrinsic processing response of titanium alloys leading to a radical shift in the manufacturing process paths including dramatic reduction or elimination of ingot breakdown enabled by an order-of-magnitude decrease in the as-cast grain size, and relaxing constraints in secondary manufacturing processes resulting in lower cost titanium.

Acknowledgements

The assistance of Ms. Linda Shepard in manuscript preparation and Mr. David Newell in drawing some figures, is greatly appreciated.

References

1. F.H. (Sam) Froes, M. Ashraf Imam and Derek Fray, Eds. "Cost Affordable Titanium" TMS, Warrendale, Pa. 2004.
2. T.E. Norgate and G. Wellwood, "The Potential Applications for Titanium Metal Powder and Their Life Cycle Impacts", JOM, Vol. 58, No. 9, Sept. 2006, pp 58-63.
3. E.H. Kraft, *Summary of Emerging Titanium Cost Reduction Technologies*, in publication (Oak Ridge National Laboratory, Nov. 2003).
4. S. Karpel, "Switched-on Titanium", *MBM*, (July 2003), p. 18.
5. P. Kirchain, *The Role of Titanium in the Automobile*, (Cambridge, MA: Camanoe Associates, July 2002).
6. F.H.(Sam) Froes, "Tenth World Titanium Conference", *Light Metal Age*, 61 (9, 10) (2003), 52-57.
7. F.H (Sam) Froes and T. Nishimura, "The Titanium Industries in Japan and USA – A Comparison", *Kinzoku (Materials Science and Technology)*, 73 (5) (2003), 9.
8. F.H. (Sam) Froes and K. (Oscar) Yu, Ti-2002, "Emerging Applications for Titanium", (paper presented at the Ti-Conference, Hamburg, Germany, July 2003).
9. Christopher Hatch and Robert Nestor "Military Makeover the Investment Casting Way" Modern Casting, Dec. 2003, pp.30-32.
10. Kevin L. Klug, Mehmet Gungor, Ibrahim Ucok, Lawrence Kramer, Christopher Hatch, Robert Spencer and Ronald Lomas, "Affordable Ti-6Al-4V Castings," *Cost-Affordable Titanium –Symposium Proceedings*, Eds, F.H. (Sam) Froes, M. Ashraf Imam, and Derek Fray TMS, Warrendale, PA., 2004, pp. 103-109.
11. Kevin L. Klug, Ibrahim Ucok, Mehmet N. Gungor, Mustafa Guclu, Lawrence S. Kramer, Wm. Troy Tack, Laurentiu Nastac, Nicholas R. Martin and Hao Dong, "Near-Net-Shape Manufacturing of Affordable Titanium Components for M777 Lightweight Howitzer: Research Summary," JOM, Vol. 56, No. 11, November. 2004, pp. 35-41.

12. Laurentiu Nastac, Mehmet N. Gungor, Ibrahim Ucok, Kevin L. Klug, and Wm. Troy Tack, Advances And Challenges in Investment Casting of Ti-6Al-4V Alloy: A Review," International Journal of Cast Metals Research, Vol. 18, No. 6, April 2006, pp. 1-22.
13. F.H. (Sam) Froes, V.S. Moxson, V. Duz and C.F. Yolton, "Titanium Powder Metallurgy in Aerospace and Automotive Components", (Paper presented at the Las Vegas MPIF Conference, June 2003).
14. F.H. (Sam) Froes, H. Friedrich, J. Kiese and D. Bergoint, "Titanium in the Family Automobile: The Cost Challenge", JOM, Vol. 56, No. 2, Feb. 2004, pp. 40-44.
15. T. Saito, "New Titanium Products via Powder Metallurgy" (Paper presented at the Ti-Conference, Hamburg, Germany, July 2003).
16. F.H.(Sam) Froes, "Advances in Titanium Metal Injection Molding", these proceedings.

LOW COST FABRICATION OF TITANIUM ALLOY COMPONENTS DIRECTLY FROM SPONGE

Dr. J. C. Withers[1], Dr. R. S. Storm, and Dr. R. O. Loutfy
[1]MER Corporation, 7960 S. Kolb Road, Tucson, AZ 85706, U.S.A.

Keywords: titanium, low cost, titanium sponge, rapid manufacturing, plasma transferred arc, casting

Abstract

The cost of a titanium alloy component is typically 10 to 30 times the cost of basic Kroll process produced sponge. In order for titanium to penetrate widespread utilization in defense, transportation and general commodity commercial application the cost of components ready for use must be dramatically reduced. Reducing the cost of primary titanium/sponge is progressing under the DARPA Initiative in Titanium. Even with low cost sponge the cost a final component via conventional processing remains too high for many applications. Alternate processing has been demonstrated to translate sponge directly into near net shape alloy components at a cost of $4 to $8 per pound over the cost of sponge. Low cost sponge and this alternate processing provides an avenue to produce truly low cost titanium components. The alternate one step processing will be described, produced components demonstrated and cost analysis illustrated.

Background

As a result of the excellent mechanical properties and corrosion resistance of titanium (Ti) alloys, combined with a density ~40% lighter than steel, designers are considering Ti alloys for a number of structural applications. Unfortunately, the cost of Ti alloys is very high compared to steel, and has been rapidly escalating. In many cases the decision as to which metal to use is based on an economic consideration. In order for titanium to penetrate widespread utilization in defense, transportation, and general commodity commercial application, the cost of components ready for use must be drastically reduced.

Contributing factors to the high cost of titanium components include the cost of the starting Ti powder combined with processing costs. Ti sponge has historically been ~$4.50-$5.00/lb. An effort is currently underway under DARPA sponsorship to reduce this cost significantly. (A team of MER and DuPont is in the scale up Phase II of this program.) However, the cost of finished components (exclusive of machining) may be 10-30X higher than this powder cost. A 50% reduction in starting powder cost by itself will not provide a reduction in component cost which is sufficient to significantly increase titanium utilization. The bulk of the finished component costs are related to downstream processing. Alternate processing has been demonstrated to translate sponge directly into near net shape alloy components at a greatly reduced cost.

A number of manufacturing processes are commonly used to produce finished Ti alloy components. The simplest may be casting of an ingot, followed by extensive machining to achieve useful shapes. The machining costs for this method can be excessive for all but the

---

[1] Author to whom any correspondence should be addressed.

simplest geometries. Casting is utilized to produce fairly complex geometries, but requires tooling which adds to component cost and has a significant lead time. Some finish machining is generally required to achieve required tolerances and surface finishes. A comparison of the costs to obtain near net shape components produced by these methods was carried out, and is shown in Table I. The data was assembled before the recent current price escalations in both sponge and processing cost. While the component costs would be higher today, their relative ranking in cost would still be consistent.

Considerable effort has been expended by universities and the National Labs to develop a rapid prototype processing based on the use of a laser as a high energy source. The basic principal is the fusion (melting) of metallic powders with a high power laser to form a fully dense pore-free deposit of the material. The laser beam and powder are both guided by 3D computer-based design and numerical control positioning software (CAD, CAM, CNC) to produce complex shape 3D components and coatings thereon without molds, forming dies, tooling or machining. The rate of build, surface finish and tolerance are dependent on the feed particle size, laser beam size, laser power and particle feed rate. A very refined microstructure is typically achieved via the inherent rapid solidification, which typically produces much better properties than wrought.

Table I: Costs for Near Net Shape Ti-6-4 Components

Processes	Assumptions	Selling Price $/lb in component form	Price/in[3]
Ingot	6" round billet, requires extensive machining	$25/lb[1]	$4
Investment casting	Quote from commercial supplier	$70	$11
Ram graphite casting	Quote from commercial supplier	$50-55	$8-9
Laser SFFF	Powder @ $60/lb with 20% overspray[2]	$129[3]	$21
PTA SFFF[4]	Volume pricing for Ti-6-4 wire @ $23/lb	$38[3]	$6
PTA SFFF	CP Ti wire @ $14/lb plus Al-V powder	$25[3]	$4
PTA SFFF	Mix Ti sponge + Al-V, roll into wire	$6-8[3, 5, 6]	$1-1.3
Sand cast A148 Stainless	Quote from commercial supplier	$6.57	$1.87

[1] Cost of fully machined component is typically $100-125/lb.
[2] The overspray in some units can often be 80% and recycle of powder has not been demonstrated to produce acceptable material which would raise price from $129/lb to $413/lb.
[3] Includes a 20% markup for profit on PTA and laser processing.
[4] PTA SFFF = plasma transferred arc solid free form fabrication
[5] Based on $4.00/lb CP sponge
[6] Potential is 2X the cost of sponge. If $2 to 3/lb sponge becomes available the cost of titanium plate for armor and other near net shapes has the potential of $4 to 8/lb which is lower cost than stainless steel.

General note: It should be understood titanium process are subject to fluctuation and quoted prices may vary. These prices were obtained mid-2004.

The down side of the laser processing is the very slow build rates and high cost of the components they can produce. Commercial laser based rapid prototyping systems are limited to small components at low volumes. This process is fully automated, so labor costs are minimal. However, deposition rates are very low, powder capture rates are very low at 10-20%, and net electrical efficiency is very low (~20%). In addition, capital costs are very high. The result of these factors is component costs of laser produced materials are much higher than standard processing (see Table I).

These difficulties can be overcome by using an electrically efficient high energy beam, and combining it with the same 3D CNC control systems used with the lasers. MER undertook

14

an extensive investigation of the options available for high energy beam rapid manufacturing. The conclusion was that the deficiencies with the laser rapid prototyping process could be addressed by using a higher energy source employed by the welding industry to melt metals. More specifically, the use of a plasma transferred arc (PTA) had the best combination of process capabilities combined with the best economics from a capital and operating cost perspective, compared to E-beam and TIG welding beams.

As shown schematically in Figure 1, the PTA torch uses a plasma to transfer an electric arc to a work piece. This results in very high deposition rates, up to 50 lb/hr using a 350 amp power supply, and higher with larger power supplies. Importantly, the PTA process can also use a micro torch at very low amperage levels (single digit amps) to deposit fine features and produce smooth surfaces. The PTA process runs at atmospheric pressure, and can be run in an inert gas chamber or shielding gas system for processing of oxidation sensitive powders. The energy levels of the torches are also sufficient to process very high melting refractory metals. Components produced using this process have a much more rapid cooling cycle than, e.g., components made by casting. As a result the mechanical properties of PTA manufactured components compare very favorably to the properties from conventional processing as shown in Table II. The capital costs of the PTA based systems are a fraction of comparable laser or E-beam systems. Based on these considerations, MER has developed a large scale PTA rapid manufacturing capability, shown in Figure 2, which can produce components with a size up to 2' x 2' x 16'. Examples of several components produced by the PTA rapid manufacturing process are shown in Figure 3. Systems are in design to directly produce thick titanium armor plate in one step at low cost.

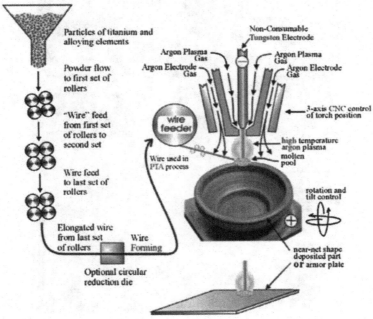

Figure 1: Schematic of plasma transferred arc (PTA) solid free form fabrication system (SFFF)

Table II: Mechanical properties of Ti-6-4 produced by various manufacturing processes

Property	Cast	Machined from billet	Powder metallurgy	Component formed from PTA manufacturing
Tensile Strength (ksi)	147	147	110-140	160
Elongation (%)	10	10	10-12	10-13
Fatigue ratio	0.54	0.54	—	0.54

Figure 2: MER Commercial scale plasma transferred arc rapid manufacturing facility

16

The cost to produce the same components described in Table I were estimated for the PTA rapid manufacturing process. Because of the low capital costs, low labor, and high efficiencies, the costs for the PTA rapid manufacturing process are considerably lower than those for casting: 45% lower than investment casting, and 25% lower than ram graphite casting. (Investment casting produces a higher quality part than ram graphite casting.) The major cost contributor for the PTA manufactured component was the Ti-6-4 feed wire, which was priced at $23/lb, vs. a component cost of $38. Today's pricing for the wire is ~2.5X higher, and MER has therefore focused on the cost of the wire feed to achieve the lowest possible Ti-6-4 component cost.

The Ti-6-4 wire is drawn from a prealloyed ingot. The cost for CP Ti wire is considerably less expensive than that of the Ti-6-4 wire. MER has demonstrated that by adding a prealloyed powder of 60%Al/40%V to the PTA melt pool, the resultant product has the same microstructure and properties as that obtained when using Ti-6-4 wire. By using a dual feed of CP Ti wire and Al-V powder, a further 35% reduction in cost was demonstrated. On a cost per volume basis, this resulted in a near net shape Ti-6-4 component cost ~2X that for sand cast stainless steel. This can be compared to a 6X premium for investment cast Ti-6-4 compared to the sand cast stainless steel.

The cost of the raw materials for the PTA process with CP Ti wire and prealloyed Al-V powder is still the dominant cost element (65-70%). MER has therefore continued to work on further reducing the cost of the wire feed stock. This has led to the demonstration of a route that does not involve melting the Ti to form the wire. Because of the high ductility of primary Ti sponge, it can be compacted by cold rolling to densities of ~70%. It is therefore possible to feed titanium sponge into sets of rollers and through successive reduction steps to cold form into a wire shape in one step as illustrated in Figure 1. Using the cold formed wire process combined with PTA rapid manufacturing, processing costs separate from powder cost are estimated to be $4-8/lb or less. With the projected cost for primary Ti sponge from the DARPA low cost powder program, the resultant cost for near net shape Ti-6-4 components can achieve parity on a cost/volume basis with cast stainless steel. This breakthrough can be the catalyst to achieve a significant expansion of Ti utilization.

Figure 3: Components produced by PTA rapid manufacturing:
Ti-6-4 sprocket carrier for the EFV, composite armor tile, ¼ scale
bimetallic penetrator, and layered Ta/Zr penetrator cubes

DEVELOPMENT OF COST EFFECTIVE BLENDED ELEMENTAL POWDER METALLURGY Ti ALLOYS

Fusheng Sun and Kuang-O (Oscar) Yu

RTI International Metals Inc.
1000 Warren Avenue, Niles, OH 44446

Keywords: Titanium alloys, Powder metallurgy, Cost Effective

Abstract

Cost effective titanium alloy compacts were produced via the powder metallurgy approach by blending elemental (BE) titanium sponge fines and master alloy. The pressed and sintered approach potentially offer relative low cost titanium hardware suitable for aerospace applications and military ground vehicle applications. Pressed and sintered BE Ti-6Al-4V compacts were characterized in terms of chemistry, microstructure and properties. Effects of hot rolling and deformation on the microstructure of BE Ti-6Al-4V compacts are reported. Processing both before and after sintering was found to have a significant impact on microstructure and properties.

Introduction

Titanium alloys are the materials of choice for aerospace, ground armor vehicles, and automotives because of their excellent strength-to-density ratio, attractive strengths at elevated temperatures and corrosion resistance. However high cost often limits their applications, so that reduced cost continues as a major goal of the titanium industry [1]. Cost reduction can come at the extraction stage or in subsequent processing. Powder metallurgy is receiving considerable attention for titanium alloys because the potential cost benefits [2-5].

Historically, two different titanium powder metallurgy approaches have been developed: 1) the hot isostatic pressing (HIP) of prealloyed powders, 2) the cold pressed and sintered process using blended elemental (BE) sponge fines. The basic advantages of cold pressed and sintered process are economic advantages resulting from lower cost powders, reduced materials losses when compared with wrought processing, and mechanical properties approaching those of wrought materials.

In the paper attempt has been made to consolidate blended titanium and master alloy powders for achieving full density Ti-6Al-4V alloy using the pressed and sintered process, and subsequent hot rolling. The sintered compacts were uniformly hot rolled at beta phase field and then hot rolled at (alpha + beta) phase field. The resulting sheets were characterized in terms of chemistry, microstructure and properties. However, this study, though limited in scope, is considered adequate to show the potential of the powder metallurgical approach to titanium mill products.

Experimental

The powders used for producing blended elemental P/M Ti-6Al-4V samples are −100 mesh Ti sponge fines and a Al-V master alloy. The chemical compositions of the titanium powder and Al-V master alloy are shown in Table I and II.

The titanium and master alloy powders were blended in conventional V-type blenders and the resultant mixture was cold –compacted. Samples were cold-compacted in a hydraulic press at a pressure of 445 MPa. This yields a green compact with a density of 84 percent of theoretical density. Sintering was performed at 2300 °F for 4 hours in a high vacuum to avoid interstitial contamination.

Table I The chemical composition of −100 mesh Ti powder (weight percent %)

Elements	O	N	H	C	Fe	Mg	Na	Cl	Ti
Wt %	0.240	0.036	221ppm	0.006	0.03	0.01	<0.01	0.043	Bal

Table II The chemical composition of Al-V master alloy (weight percent %)

Elements	Al	V	O	N
Wt %	43.17	56.37	0.185	0.037

The sintered compacts were subsequently processed to achieve full dense by hot cross rolling at 1950 °F with a reduction of 65 %, followed by hot cross rolling at 1750 °F with a further reduction of 65%. The resultant Ti-6Al-4V sheets with a thickness of 0.08" were flattened and mill annealed. After grounding and pickling, tensile samples were prepared from the sheets with either longitudinal direction or transverse direction. Tensile testing was performed with a strain rate of approximately 0.01 in/in/min. Metallographic and fractographic characterization was performed using optical and scanning electron microscope techniques.

Results and Discussions

Sintering Characteristics

The microstructure of the pressed and sintered BE Ti-6Al-4V material is shown in Figure 1. The microstructure consists of coarse lenticular alpha and black porous area. The volume fraction of the Ti phase and the porous area within the sintered compact was calculated for this sample: approximately 93.2% Ti phase and 6.8% pores. The pores are about 10-30 µm in diameter and there is a tendency for these pores to locate near the grain boundaries and at triple joints. The sintered BE Ti-6Al-4V density is dependent upon processing parameters and titanium particle attributes such as compacting pressure, sintering temperature and time, particle size and chlorine content in titanium powder [2-5]. Previous study [4] showed that a 95% dense BE Ti-6Al-4V material in the as-sintered condition of 2300 °F/ 4 hours was obtained when a cold compacting pressure of 620 MPa was used. In another study [2], a sintered density of 96% had obtained in BE titanium under a condition of 2370 °F for 4 hours when a cold compacting pressure of 686 MPa was used. Compared to the previous studies, the sintered density of 93.2% in the BE Ti-6Al-4V is reasonable when considering the compacting pressure of 445 MPa used in the present study.

The SEM images of the pressed and sintered BE Ti-6Al-4V material are shown in Figure 2. The microstructural features (Figure 2a) appear to be the same as those observed in optical microscope (Figure 1) with exception that some titanium particles were observed in pores. The higher magnification SEM image in Figure 2b shows that the microstructure consisted of lenticular alpha (grey) and beta phase (white) virtually uniform from the edge to the middle of the compact. A titanium particle was also observed within the pore as shown in Figure 2b.

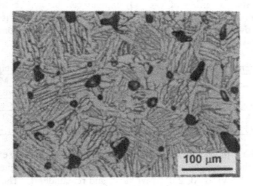

Figure 1. Microstructure of the pressed and sintered BE Ti-6Al-4V material.

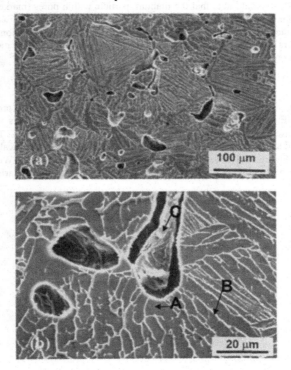

Figure 2. SEM images showing the microstructure of the pressed and sintered
BE Ti-6Al-4V material.

21

Composition

The chemical analysis of the sintered compacts is illustrated in Table III. The Al, V, Fe, C, N, and H contents in the sintered BE Ti-6Al-4V compacts are within the wrought specifications (AMS 4911) with the exception of O_2. O_2 content can be held below the 0.2% wrought specification if titanium powders with relatively lower oxygen content is used. Compared with the input titanium powder as shown in Table I, the pick-up of oxygen content during sintering is in the range of 0.06 weight percent.

Table III The chemical composition of the sintered BE Ti-6Al-4V (weight percent %)

Elements	Al	V	O	N	H	C	Fe	Mg	Na	Cl	Ti
Sample 1	5.52	4.34	0.290	0.041	0.0005	0.036	0.04	<0.01	<0.01	<0.01	89.71
Sample 2	5.53	4.34	0.281	0.042	0.0075	0.044	0.04	<0.01	<0.01	<0.01	89.69
Average	5.53	4.34	0.286	0.042	0.0040	0.040	0.04	<0.01	<0.01	<0.01	89.70
AMS 4911	5.50-6.75	3.50-4.50	0.200	0.050	0.0150	0.080	0.30	-	-	-	-

The alpha or beta phase in the sintered BE Ti-6Al-4V compacts was also examined using EDAX. The alpha phase (the grey phase marked A in Figure 2b) has higher Al content, with a composition of $7.0^{\pm 0.8}$ wt % Al, $2.0^{\pm 0.5}$ wt % V, and remainder Ti. The beta phase (the white phase marked B in Figure 2b) shows a higher V content of $8.0^{\pm 0.8}$ wt % and a lower Al content of $4.5^{\pm 0.5}$ wt %. It should note that the titanium particle within pores (marked C in Figure 2b) has a composition of $6.0^{\pm 0.8}$ wt % Al and $3.8^{\pm 0.5}$ wt % V, within a range between alpha and beta phase. This indicated that the sintering process occurred as interdiffusion process between titanium powders and Al-V master alloy powders followed by recrystallization process. Those particles within pores were not completely recrystallized.

Hot Rolling

Hot rolling was applied to demonstrate the feasibility for producing flat products (plates and sheets) from BE Ti-6Al-4V alloy. The sintered BE Ti-6Al-4V compacts were processed to achieve full dense by hot cross rolling at 1950 °F with a reduction of 65 %. The microstructure of the hot rolled sheets consisted of a large beta grain structure, with a typical beta grain size of about 200 μm (Figure 3). The beta grains contain a high volume fraction of Widmanstätten alpha plates of high aspect ratio. Some voids were observed along the grain boundaries or triple joints.

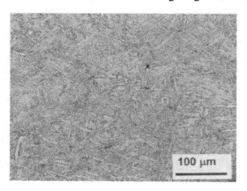

Figure 3. Microstructure of the BE Ti-6Al-4V material hot rolled at 1950 °F.

Figure 4 shows the SEM image of the BE Ti-6Al-4V alloy hot rolled at 1950 °F by a reduction of 65%. As can be seen in the figure, the microstructure is of a Widmanstätten type with high aspect ratio alpha plates. Although microvoids with a diameter of about 1-3 μm were occasionally observed, the density of the hot rolled BE Ti-6Al-4V is 99+% of full density.

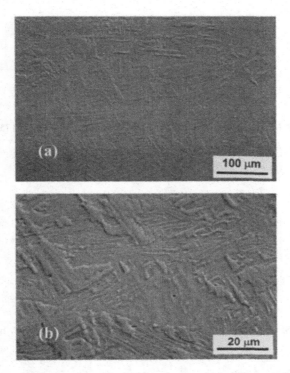

Figure 4. SEM images showing the microstructure of the hot rolled
BE Ti-6Al-4V material at 1950 °F.

In order to achieve desired levels of mechanical properties, hot cross rolling at (alpha + beta) phase field was subsequently performed for the above hot rolled sheets. Figure 5 shows the typical optical microstructure with the longitudinal direction of the sheets after hot rolled at 1750 °F with a reduction 65%. Due to the large (alpha + beta) hot rolling, the microstructure of the sheets generally consisted of a uniform, fine and globular (alpha + beta) microstructure (Figure 5). As can be seen from the SEM image in Figure 6, a 99+% dense BE Ti-6Al-4V alloy sheet was obtained. Details of the microstructural features are revealed by high magnification SEM micrographs as shown in Figure 7. It can be seen that the (alpha + beta) hot-rolled microstructure consists of a very high volume fraction of primary-alpha globular (or elongated) gains (Figure 7a). The grain size of primary-alpha globular gains was in a range of 3-5 μm (Figure 7b). Some round-shaped microvoids were occasionally observed in the microstructure.

23

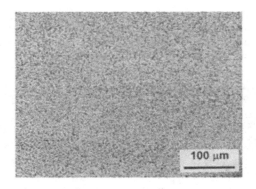

Figure 5. Microstructure of the BE Ti-6Al-4V material hot rolled at 1750 °F.

Figure 6. SEM images showing the microstructure of the BE Ti-6Al-4V material hot rolled at 1750 °F.

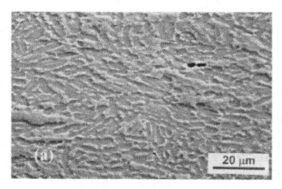

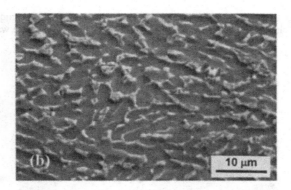

Figure 7. High magnification SEM images showing the microstructure
of the BE Ti-6Al-4V material hot rolled at 1750 °F.

Mechanical Properties

The (alpha + beta) hot-rolled BE Ti-6Al-4V sheets with a thickness of 0.080" were flattened and annealed at 1450 °F for 1 hour, and then grounded and pickled for mechanical properties testing. Room temperature tensile properties of the hot rolled BE Ti-6Al-4V sheets with both the longitudinal and transverse directions are shown in Table IV. It is obvious that the tensile properties of the hot rolled BE Ti-6Al-4V sheets exceed the mechanical property requirements of AMS 4911J (wrought Ti-6Al-4V annealed plates and sheets). The BE Ti-6Al-4V sheet material of this study compares favorably in ultimate tensile strength, yield strength, and ductility to that of conventional wrought Ti-6Al-4V sheet, particularly in the tensile strengths. Since hot rolling was accomplished in a uniform manner, that is, with the sheet being rotated 90° after each pass, directionality effects were unexpected.

Table IV Tensile properties of 0.08" sheet hot rolled from P/M Ti-6Al-4V compacts

Specimen		UTS		YS		El., %
		ksi	MPa	ksi	MPa	
BE Ti-64 sheet	1L*	160	1104	140	966	17.6
	2L*	160	1104	139	959	19.6
	1T*	158	1090	136	938	19.6
	2T*	159	1097	137	945	11.8
AMS 4911J		134	925	126	869	10

* 1L and 2L with the longitudinal direction; 1T and 2T with the transverse direction.

Fractograhpy

The fractured surface of a typical BE Ti-6Al-4V tensile sample is shown in Figure 8. The predominant fracture modes were similar to those found in corresponding conventional wrought material. The ductile nature of the fracture in the BE Ti-6Al-4V material was shown by the presence of dimples (Figure 8a). Some large pores on the fractured surface were observed (Figure 8b). Those large pores may result from microvoids (as shown in Figure 7a) found after hot rolling. Attempt was made to identify the presence of trace chlorine and sodium (salts) in the

25

dimples and large pores on the fractured surfaces using the EDAX technique. But no the presence of salts (chlorine and sodium) was found. The fresh fractured surface was virtually very "clean".

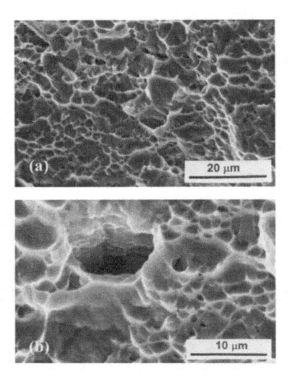

Figure 8. Scanning electron micrographs of the fracture surface of a BE Ti-6Al-4V tensile specimen.

Cost effective elemental blended Ti approach

The tensile properties of the hot rolled BE Ti-6Al-4V sheets are quite encouraging. The good combination of tensile strength and ductility in the BE Ti-6Al-4V material is attributed to both its increased density and relatively fine equiaxed microstructure. It is well known that an increase in the density of P/M titanium alloys leads to increased strength and ductility. The hot rolled BE Ti-6Al-4V material in the present study exhibits a 99+ % of full density with a uniform microstructure. This is due to the low chlorine contents (on the order of 0.05 weight percent) of the input titanium powders. Some rounded microvoids with a diameter of 1-3 μm were observed in the hot rolled sheets. However no significant deleterious effects of those rounded microvoids on room temperature tensile properties was found. The microvoids may result from residual chlorine content or severe deformation, which are subject to further study.

The density of P/M titanium alloys is significantly dependent upon both P/M process parameters and chemical compositions (particularly chlorine contents). Numerous studies [3-5] have shown that residual chlorine has a strong effect on the density of P/M Ti-6Al-4V material. The deleterious effects of chlorine on the density of hot pressed P/M Ti-6Al-4V showed that the residual chlorine (on the order of 0.15 weight percent) related large pores were impossible to be eliminated after hot pressing [5]. Lower chlorine content titanium powder is required and is one of the major factors for achieving a good performance BE Ti-6Al-4V material.

The attractive properties of the BE Ti-6Al-4V material are also significant because the blended elemental approach offers economic advantages for parts amenable to P/M processing. The cost effective BE Ti-6Al-4V material is due to the relatively low-cost powder used and the simple pressed and sintered approach. The combination of the attractive mechanical properties and cost-effective advantages should significantly increase the applications of BE Ti-6Al-4V material.

Conclusions

The feasibility of using the pressed and sintered bended elemental Ti-6Al-4V approach in corporation with mill processing to produce cost effective flat Ti-6Al-4V products (plates or sheets) has been demonstrated in this study. A 99+% of full dense BE Ti-6Al-4V sheet material has been developed, with a uniform fine equiaxed microstructure. The room temperature tensile properties of the BE Ti-6Al-4V sheet material compare favorably to those of conventional wrought sheet. The tensile strengths and ductility exceed the mechanical property requirements of AMS 4911J (wrought Ti-6Al-4V annealed plates and sheets).

References

1. F. H. Froes, K. Yu, and T. Nishimura, "Developing Applications for Titanium," *Cost-Affordable Titanium*, ed F. H. Froes, M Ashraf Imam, and Derek Fray (Warrendale, PA: The Minerals, Metals & Materials Society, TMS, 2004), 19-26.

2. T. Saito, "New Titanium Products via Powder Metallurgy Process," *Ti-2003 Science and Technology*, ed. G. Lütjering and J. Albrecht (Frankfurt, Germany: Deutsche Gesellschaft für Materialkunde DGM, 2003), 399-410.

3. D.A. Duz et al., "Powder Metallurgy Ti-6Al-4V Components Produced from Low Cost Blended Elemental Powders by Hot Pressing," *Cost-Affordable Titanium*, ed F. H. Froes, M Ashraf Imam, and Derek Fray (Warrendale, PA: The Minerals, Metals & Materials Society, TMS, 2004), 145-150.

4. P. J. Andersen and P. C. Eloff, "Development of Higher Performance Blended Elemental Powder Metallurgy Ti Alloys," *Powder Metallurgy of Titanium Alloys*, ed F. H. Froes and J. E. Smugeresky (Warrendale, PA: The Metallurgical Society of AIME, 1980), 175-187.

5. R. R. Boyer, J. E. Magnuson, and J. W. Tripp, "Characterization of Pressed and Sintered Ti-6Al-4V powders," *Powder Metallurgy of Titanium Alloys*, ed F. H. Froes and J. E. Smugeresky (Warrendale, PA: The Metallurgical Society of AIME, 1980), 203-216.

A Novel Process for Cost Effective Production of TiAl based Titanium Alloy Powders

G.Adam[1], D. L. Zhang[2], Jing Liang[1]
[1]Titanox Development Limited-New Zealand
[2]The University of Waikato -New Zealand

Abstract

Titanium alloys are among the most favourable advanced materials of our time because of their excellent combination of specific mechanical properties, outstanding corrosion resistances and reasonally good high temperature oxidation resistance. Among them, titanium aluminides (TiAl and Ti_3Al) have been proven to have the potential to be used in a wide range of industrial applications. However, the production of titanium aluminides is still not cost-intensive, and their mechanical properties and the effects of further alloying, are still under investigation. In conjunction with the University of Waikato, Titanox Development Limited in New Zealand has successfully developed a novel process to produce TiAl alloy powders with very fine particle sizes cost effectively. The cost of the powders produced using this process is only a very small fraction of the international market prices of such powders. This technology is mainly characterised by process steps including mechanical milling, heat treatment, physical separation, and chemical reduction. Titanium aluminide based powders with composition range from Ti_3Al to TiAl with high purity have been produced and successfully used in making coatings and near net shaped powder metallurgy parts. The process has been disclosed in a PCT (Patent Corporation Treaty) application which was approved in 2004 [1], and the related patent applications either have been approved or are being filed in different countries.

Introduction

Advanced materials are key to many technological advancements, and enhanced structural materials are particularly vital to aerospace system improvements where the extremely demanding conditions require that the new materials are stronger, stiffer, lighter, and being able to survive at higher temperatures. These advanced materials are often tailored or engineered to achieve the properties required for a given application by use of composite concepts [2-6]. An attractive class of "advanced materials" is the intermetallic compounds [7]. Intermetallic compounds are phases that occur in the central parts of the phase diagrams between two or more metals, and may have very specific compositions or a range of compositions [7-11]. Titanium and its alloys offer physical advantages over their material competitors to use in many industrial applications. Over the last decade the use of titanium alloy in non-aerospace and non-military applications such as automobile industries and any other industrial applications are still limited due to the high manufacturing cost associated with the expensiveness of raw materials and processing difficulties.

This paper is to describe and discuss the major results of a study on the new process which has been developed by Titanox Development Ltd and used to produce TiAl based alloy powders. The objectives of this study is to determine the feasibility of producing TiAl based alloy powders by processing TiO_2 and Al powders using the novel process and to establish essential relationships between processing conditions and powder characteristics associated with the process.

Experimental Procedure

The experimental procedure used in this study follows the schematic diagram shown in Fig.1. The weight of starting TiO_2 and Al powders in the starting powder mixture was controlled according to the following nominal chemical equation:

$$3TiO_2 + 4Al \longrightarrow 3Ti + 2Al_2O_3 \qquad (1)$$

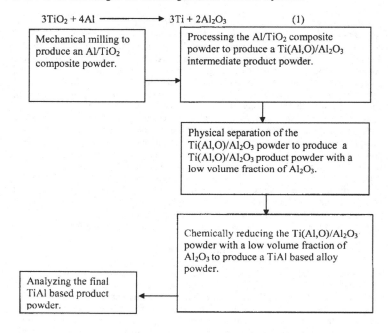

Figure1. Schematic diagram of the experimental procedure used to produce the TiAl based alloy powder with fine particles.

Mechanical milling of a mixture of TiO_2 and Al powders was performed for 2 hours using discus-milling machine to produce a TiO_2/Al composite powder. The mechanical milling was followed by heat-treatment of the TiO_2/Al composite powder at 900°C for 2 hours and then sintering of the heat treated powder at 1600°C for 4 hours using tube furnaces. The heat treatment process turned the TiO_2/Al composite powder into $Ti(Al,O)/Al_2O_3$ composite powder, and the sintering process consolidates the composite powder into bulk composite pieces [12]. The milling, heat-treatment and sintering processes were all performed in an inert argon gas environment. The

30

heat treated powder and sintered bulk pieces were subsequently crushed using the
same high energy discus milling machine to produce a powder for the later physical
separation step. The powder from the crushing step was wet milled in the presence of
a suitable surfactant using a planetary mill, and then separated using a sedimentation
column. The final chemical reduction of the titanium rich powder from the physical
separation step using a suitable reducing agent was performed to produce pure TiAl
based alloy powder with very fine particle sizes.

Results

Separation of the Fine Ti (Al, O)/Al$_2$O$_3$ Powder Using a Sedimentation Column

The XRD pattern of the starting fine Ti(Al,O)/Al$_2$O$_3$ powder produced by crushing the
as-heat treated Ti(Al,O)/Al$_2$O$_3$ composite powder is shown in Fig. 2. The starting
XRD pattern revealed Ti(Al,O) and Al$_2$O$_3$ as the major phases and TiO as a minor
phase. The Al and O in the bracket represent the Al and O dissolved in the solid
solution of titanium matrix. Fig. 3 shows the cross sections and morphology of the
powder particles in the starting fine Ti(Al,O)/Al$_2$O$_3$ powder. Ti(Al,O) particles (bright
particles in Fig. 3(a)) and Al$_2$O$_3$ particles (grey particles in Fig. 3(a)) had similar sizes
and morphology. As shown in Fig. 3(b), the Ti(Al,O) and Al$_2$O$_3$ particles were of
equiaxed shapes and slightly agglomerated.

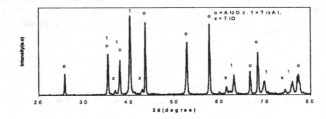

Figure 2.XRD pattern of the starting fine Ti(Al,O)/Al$_2$O$_3$ powder produced by
crushing the as heat-treated Ti(Al,O)/Al$_2$O$_3$ composite powder.

31

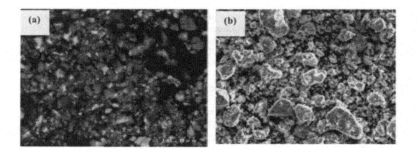

Figure 3. (a) SEM backscattered electron micrograph showing the cross sections of the powder particles in the fine Ti(Al,O)/Al$_2$O$_3$ powder, (b) SEM secondary electron micrograph showing particle morphology of the powder.

The fine Ti(Al,O)/Al$_2$O$_3$ powder was separated using sedimentation column, after it was wet milled in the presence of a suitable surfactant. As shown in Fig. 4, the XRD patterns of the powders collected from suspension and sediment zones of the column (after drying out) show that the powders from two zones of the column do not have any significant difference in the phase composition. This confirms that separation of the Ti(Al,O) and Al$_2$O$_3$ particles in the fine Ti(Al,O)/Al$_2$O$_3$ powder was not successfully achieved.

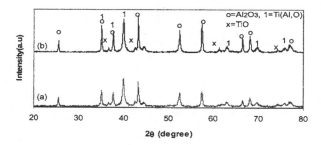

Figure 4. XRD results of the powders collected from different zones of the sedimentation column after the separation of the fine Ti(Al,O)/Al$_2$O$_3$ powder, (a) the powder from the suspended particles zone, and (b) the powder from the sediment zone.

Separation of the Coarse Ti(Al,O)/Al$_2$O$_3$ Powder Using a Sedimentation Column

Bulk Ti(Al,O)/Al$_2$O$_3$ composite material was produced from the as-heat treated Ti(Al,O)/Al$_2$O$_3$ composite powder was produced by pressureless sintering of the composite powder for 4 hours at 1600°C. Crushing of the sintered pieces resulted in formation of a coarse Ti(Al,O)/Al$_2$O$_3$ powder with the same phase compostion as the

32

fine Ti(Al,O)/Al$_2$O$_3$ powder. This coarse powder was then wet milled in the presence of the a suitable surfactant. Fig.5 shows the XRD patterns of the powders collected from different zones of the sedimentation column. From the XRD analysis of the two powders collected from the sedimentation column, it was clear that the powder collected from the suspension zone showed a large increase in the fraction of Ti(Al,O) particles as compared with that in the starting powder, while the powder colelcted from the sediment zone of the column showed a signficant decrease in the fraction of Ti(Al,O) particles in the starting powder. The reverse is true for the fraction of Al$_2$O$_3$ particles in the powder. This differentiation shows that the use of the coarse powder in conjunction with the sedimentation column was very effective in separating the Ti(Al,O)/Al$_2$O$_3$ powder into a Ti(Al,O) rich and Al$_2$O$_3$ rich powders.

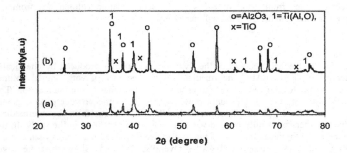

Figure 5. XRD patterns of the powders collected from different zones of the sedimentation column after the separation of the coarse Ti(Al,O)/Al$_2$O$_3$ powder (a) the powder from suspension zone, and (b) the powder from the sediment zone.

Fig. 6 shows the morphology of the particles in the powders from the suspension and sediment particles zones respectively. It appeared that the coarse particles collected from the sediment zone powder were more rounded than the fine particles collected from the suspension zone. Further chracterisation of the powders using scanning electron microscopy and energy dispersive X-ray (EDX) spectrometer clearly confirmed that the separation of the coarse Ti(Al,O)/Al$_2$O$_3$ powder into a Ti(Al,O) rich powder and an Al$_2$O$_3$ rich powder through the column sedimentation has been successfully achieved.

Figure 6. SEM seconary electron micrographs showing the particles morphology of powder particles from different zones of the sedimentation column: (a) from the suspended zone, and (b) from the sediment zone.

Fig. 7 Shows the XRD pattern of the final product powder produced by chemically reducing the Ti(Al,O)/Al₂O₃ intermediate product powder of with limited volume fraction of Al₂O₃ followed by separating the by-product phase from the reaction. The XRD analysis of the final product powder shows that the powder predominately consists of TiAl the phase. This means that the Al₂O₃ particles in the intermediate product powders were fully reacted with the reducing agent used in the study. In the meantime, the chemical analysis of the final powder showed that the oxygen content of the powder was lower than 0.4wt%. Overally, this study shows that the novel process developed by the Titanox Development Ltd is highly capable of producing TiAl based alloy powders. This a cost effective process for producing TiAl based alloy powders due to the low cost of starting materials (Al and TiO₂ powders) and the use of equipment which do not require high capital investment to construct.

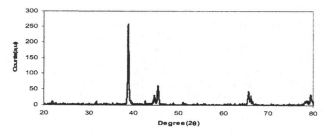

Figure 7. XRD pattern of the final product powder of pure TiAl based alloy powder produced by chemically reducing the intermediate Ti(Al,O)/Al₂O₃ powder with a limited volume fraction of Al₂O₃.

The particles morphology and size of the final product of TiAl powder are shown in Fig. 8. The SEM micrograph shows very fine particles of equiaxed shapes.

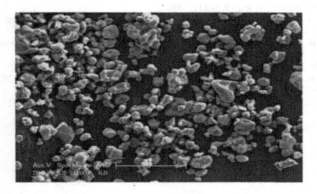

Figure 8. SEM micrograph showing the particles morphology and size of the final powder product of TiAl alloy.

Conclusions

1. Crushing of the fine structured Ti(Al,O)/Al$_2$O$_3$ composite powder and sintered bulk Ti(Al,O)/Al$_2$O$_3$ composite results in fine and coarse Ti(Al,O)/Al$_2$O$_3$ powders with fine discrete Ti(Al,O) and Al$_2$O$_3$ particles, respectively.

2. The use of the fine Ti(Al,O)/Al$_2$O$_3$ powder in combination of a sedimentation column does not result in an effective separtion of the powder, while the use of the coarse Ti(Al,O)/Al$_2$O$_3$ powder and the sedimentation column was very effective in physically separating the Ti(Al,O)/Al$_2$O$_3$ powder into an intermediate product powder with a high volume fraction of Ti(Al,O) particles and an intermediate product powder with high volume fraction of Al$_2$O$_3$ particles.

3. The final product powder of TiAl based alloy was produced by performing a novel chemical processing of the Ti(Al,O)/Al$_2$O$_3$ with a low volume fraction of Al$_2$O$_3$ particles.

4. The novel process developed results in a significantly lower cost for producing TiAl based powders because of the low cost of the starting Al and TiO$_2$ materials and the use of equipment which do not require very high capital investment to construct.

References

1. D. L. Zhang, G. Adam and J. Liang, "A Separation Process", PCT Patent Application No. PCT/NZ2003/00159, Approved in 2004
2. A. R. C. Westwood, *Met. Trans. B*. **19B**, p. 155, 1988.
3. F. H. Froes, *P/M in Aerospace and Defense Technologies*, p. 23, 1990.

4. Idem, *Mater. Design*, **10**, p. 110, 1990.
5. F. H. Froes, *P/M-Key to Advanced Materials Technology*, (Proceedings of ASM International sponsored meeting), 1990.
6. F. H. Froes and C. Suryanarayana, *Proceedings of Workshop on Advanced Materials*, 1989.
7. F. H. Froes, *Space Age Metals Technology*, p. 1, 1988.
8. H. Jeong, D. K. Hsu, R. E. Shannon and P. K. Liaw, *Metall. Trans, A* **25A**, 799 (1994).
9. H. Jeong, D. K. Hsu, R. E. Shannon and P. K. Liaw, *Metall. Trans, A* **25A**, 811 (1994).
10. E.Vagaggaini, J. -M. Domergue, and A. G. Evans. To be published.
11. D. B. Marshall and B. N. Cox, *Acta metall*, **35**, 2607 (1987).
12. D. L. Zhang, Z. H. Cai and G. Adam, JOM, February 2004, p.53-56.

FLOWFORMED Ti-6Al-4V

Mehmet N. Gungor, Lawrence S. Kramer, Hao Dong, Ibrahim Ucok, Wm. Troy Tack

Concurrent Technologies Corporation
100 CTC Drive; Johnstown, PA 15904, USA

Keywords: Flowforming, Ti-6Al-4V, Extrusion, Microstructure, Tensile, Fatigue

Abstract

This paper describes the results of an effort to develop Ti-6Al-4V seamless tubes for structural applications by the flowforming process. Flowforming is a cost-effective, cold metal forming process used for producing rotationally symmetrical, hollow components requiring a minimal or no final-machining operation. Due to severe cold work, the flowforming process produces a very fine microstructure that yields to enhanced mechanical properties. Flowforming process also generates high residual stresses requiring stress-relief treatments. This paper provides a brief discussion on the economics of flowformed material, microstructural characterization, texture development, and an assessment of tensile and fatigue properties.

Introduction

Flowforming, performed at room temperature, produces rotationally symmetrical and hollow components with high dimensional tolerances. Flowformed material is produced by application of severe compression onto the outer surface of a rotating cylindrical preform by rotating rollers. The preform material is held by a rotating cylindrical mandrel. Under high compressive stresses, material flows in the axial direction of the mandrel and produces a tubular product with reduced wall thickness and increased length. In operation, a cylindrical preform, with a pre-calculated wall thickness, length, and inside and outside diameter, is placed on the hardened steel mandrel. Forming rollers with specific profiles are set at precise distances from each other and the mandrel. When the flowforming machine is activated, depending on the configuration of the machine, the rollers exert compressive stresses as high as 75 ksi, thereby causing the material to deform and flow [1]. The output material is a tube whose material has been significantly cold worked and dimensionally controlled by the process to produce a uniform wall thickness with straight and concentric product features. During processing, adiabatic heat is generated by friction between the rollers and the work piece, and from plastic deformation of material. Generated heat is usually dissipated by application of a fluid coolant as shown in Figure 1 [1]. Flowforming methods include forward flowforming and reverse flowforming [2]. Forward flowforming is utilized for components with one closed or partially-closed end such as a closed bottom or partially closed cylinder. Reverse flowforming is used to produce a component with two open ends such as a tube. In either method, the finished part is thinner and longer than the original preform, although the volume of material remains unchanged. The final tube thickness is determined by the roller-to-mandrel spacing. Using multiple rollers that may be staggered and set at different gaps, it is possible to make multiple reductions in a single, economical machine pass [1].

Figure 1. Flowforming process showing rotating mandrel with a work piece being flowformed by three rotating and compressing rollers. Coolant fluid is used to remove generated heat. Courtesy of PMF Industries, Williamsport, PA [1].

Flowforming can be applied to many conventional ferrous and nonferrous metals and alloys. Two key considerations for preform material include; (i) grain size and (ii) ability to sustain excessive compressive strain without a failure (i.e., ductile enough to plasticize and flow without cracking). Finer grain materials can flow better, which is why preforms are usually machined from extruded or forged stock. More ductile materials can better absorb deformation without cracking. The preform geometry is also another key consideration and is usually proprietary to flowforming manufacturers. Furthermore, roller geometry, rotation speed (rollers and mandrel) and feed rate are important processing parameters. Another clear technical advantage of the flowforming process compared to extrusion, at least for titanium, is that the flowforming process can yield thin wall tube materials with improved mechanical properties. Flowformed materials have high geometrical tolerances and are more straight and concentric than those produced by other processes such as extrusion [3, 4].

Materials Efficiency

Flowforming is a highly efficient materials processing route to produce tubular products, compared to other processes such as extrusion and rotary piercing. The efficiency is measured by the amount of material wasted to attain the final tubular product in final dimensions. Conventionally, extrusion is used as a means to manufacture Ti-6Al-4V seamless tubes [5–9]. Extrusion is performed at elevated temperatures, either in the β phase field (e.g., β extrusions) or the α+β phase field (e.g., α+β extrusions). Seamless tubes can also be manufactured by the rotary piercing process in which a heated round billet is pierced through its center by a mandrel, and the billet is drawn length-wise over the mandrel by the pressure of rotating rolls on the external surface [10, 11]. Extruded Ti-6Al-4V tubes are produced with a varying thickness of additional material, on both, internal and external surfaces. Oversized extruded materials require post-processing machining in both inner and outer surfaces while flowformed Ti-6Al-4V tubes require minor or no post-processing machining. However, flowforming preforms are often made by machining of oversized extruded tubes which contributes to material waste.

Beta, α+β extrusions and flowforming approaches were utilized to produce seamless Ti-6Al-4V tubes to study material efficiency along with the other study objectives. Most of the tubes were made by utilizing the standard grade Titanium Ti-6Al-4V. The finished tube had a final dimensional specification of 14.3cm ID, 0.36cm wall thickness and 170cm length. The tube manufacturers were asked to produce tubes that were closest to the final desired size to minimize

38

final machining costs. Beta extruded tubes were produced with a relatively thicker wall and while α+β extruded tubes were produced with a minimal extra wall thickness. Flowformed tubes were either slightly machined or not machined at all. Material loss per unit length due to machining for each type of tube was calculated. The results showed that the material waste was the minimum for the flowformed tubes without considering material loss due to preform machining.

Materials Processing

For this study Ti-6Al-4V tubular materials were produced by various processes as listed in Table I. Materials suppliers were asked to provide as-processed and final shaped tubes. Flowformed materials were provided by two U.S. suppliers. An example of a flowformed tube is displayed in Figure 2.

Figure 2. Flowformed Ti-6Al-4V structural tube (about 2 meters long).

Table I. Materials Produced and Evaluated

Ti-6Al-4V Grade	Manufacturing	Annealing	Stress Relief
Grade 5	β Extrusion	705–760 °C/2 h	No
Grade 5	α+β Extrusion	705 °C/2 h	No
Grade 5	Flowformed-1*	No	540 °C x 4 h
Grade 5	Flowformed-2*	No	540 °C x 4 h

*1 and 2 indicate two different suppliers.

Microstructure Characterization

The preform microstructure of Flowformed-1 tube materials was not disclosed by the participating vendor thus it was not evaluated. Typical microstructures of the preform material that was used to make the Flowformed-2 tube (Figure 2) are shown in Figure 3a and 3b, respectively. The preform was a α+β lightly extruded material. Figure 3c gives a typical microstructure of β-extruded tube. All micrographs were taken in the direction parallel to the longitudinal direction of the tube materials. The microstructure of the α+β extruded preform material (Figure 3a) consists of large primary α grains (light) and a lamellar structure, mixture of α plates and transformed β (dark). Figure 3b clearly shows that the flowformed material has an extremely fine "fiberlike" microstructure where primary α (light) and β (dark) phases are severely elongated due to extensive deformation during flowforming. Beta-extruded material microstructure displayed in Figure 3c shows a much coarser microstructure that consists of α at prior β grain boundaries and transformed β that consists of acicular α delineated by β.

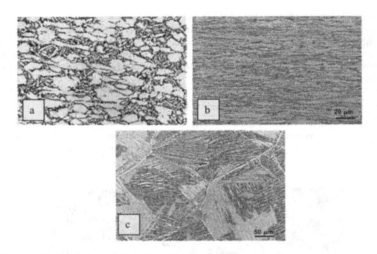

Figure 3. Optical micrographs of Ti-6Al-4V tube: (a) α+β lightly-extruded preform, (b) flowformed α+β, and (c) β extruded materials (all in the longitudinal direction).

Material microstructures were further studied using 200 kV JEOL 2010 and Philips CM20 analytical Transmission Electron Microscopes (TEMs). TEM images of the flowformed material are shown in Figure 4. Attached scanning units on these instruments allow Scanning Transmission Electron Microscopy (STEM) and thin window EDS detector systems provide elemental analysis.

The Orientation Imaging Microscopy (OIM) study was performed on a JEOL 6380 Scanning Electron Microscope (SEM) using EDAX/TSL OIM hardware and software. The 6380 resolution was not adequate to resolve the fine grain sizes and subgrains of flowformed samples. The samples were examined with a Philips XL30 FEG (field emission gun)-SEM. This method was used to collect OIM data on the stress-relieved sample. Attempts to collect OIM data from the as-flowformed sample were not successful. OIM uses electron back-scattered diffraction patterns to determine crystal direction. Heavily distorted grains have a great spread of crystal orientations and produce patterns which are not clear enough for identification by the OIM methods.

TEM and OIM of as-Flowformed Material: The micrographs showed a feature thickness of about 200 nm. The microstructure exhibits severe plastic deformation effects (Figure 4a). Regions with high dislocation densities predominate. Two grain morphologies are evident: long, thin grains (~ 100nm by 1μm) and small, equiaxed grains (~100nm–200nm diameter). Pileups of parallel dislocations, moiré' fringes from overlapping fine grains and contrast from wavy dislocations may be seen within the larger grains, but are not easily distinguished at low magnification (relative to the small scale of the microstructure) and at such a high dislocation density. The small grains (e.g., those at the upper left of Figure 4a have relatively lower dislocation densities and a few appear undeformed, indicative of a small amount of dynamic recrystallization or the formation of recrystallization nuclei. Some recovery or polygonization effects are seen in the upper central region of Figure 4a, with subgrains enclosed by low angle boundaries.

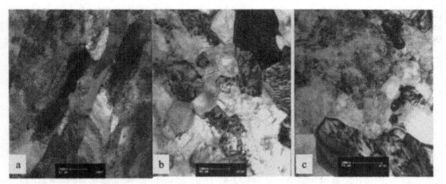

Figure 4. TEM images of the as-flowformed (a) and (b), and the heat treated Ti-6Al-4V (c).

TEM and OIM of Stress Relieved Material: The micrographs show a grain size range of about 50nm–500nm. Again, there is evidence of severe plastic deformation effects (Figure 4b) and a high dislocation density; however, there is also evidence of recovery and recrystallization. The long, thin grains seen in the as-flowformed condition are replaced by equiaxed grains. Many of the small grains (e.g., those in the left center of Figure 4b), appear undeformed, and are probably recrystallized. Polygonization effects are seen in the lower right region (Figure 4c) with subgrains enclosed by low angle boundaries, with visible contrast from the individual dislocations forming the low angle boundaries.

OIM analysis of the stress relieved material was also a challenge. Some of the results of the analysis of this dataset are shown in Figure 5. The black arrow is a marker to show the data range used for the plot of Figure 5d, which shows the misorientation range within the most heavily deformed grain. This represents a spread of about $10°/\mu m$, which indicates that it is still heavily deformed, even after the stress relief treatment. The profile shows a nearly continuously increasing misorientation within the grain over the range measured.

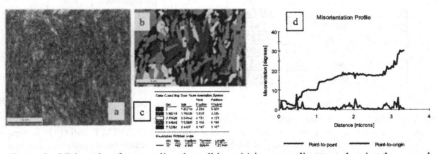

Figure 5. OIM results of stress relieved condition: (a) image quality map, showing the general microstructure, full scale, (b) enlargement of (a), displayed as grain orientation spread, showing the amount of lattice distortion within grains, (c) key, and (d) misorientation range plot for the trace in (b).

Texture Analysis: The OIM data were used to calculate orientation distribution functions, which were then used to generate the contour plot pole figures shown in Figure 6. The pole figures are aligned just as the OIM images are aligned for the transverse section samples: the lateral directions (labeled TD) are through-thickness and the vertical direction of the pole figures

41

(labeled RD) are aligned with the tangential direction of the tubes. Pole figures were used to show the alignment of poles, or plane normals, for the two major slip planes in Ti-6Al-4V, i.e., the basal (0002) and prism (10-10).

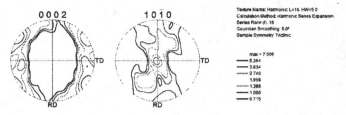

Figure 6. Pole figures of heat treated Ti-6Al-4V (with prism poles parallel to the tube axis).

Mechanical Properties

Tensile Properties: Blanks from the tube materials were extracted in the axial direction of the tube and further machined into flat tensile specimens and tested in accordance with ASTM E 8 [12]. The tensile test results are provided in Table II. Each data point is the mean of three tests. The results showed that both flowformed materials have superior properties over the extruded α+β preform. The properties of the Flowformed-2 material had an edge over the Flowformed-1 material while both materials were superior to the β extrusion. This also illustrates the difficulty in extruding thin walled titanium, in that for challenging extrusions like this, higher preheat temperatures are required to reduce the extrusion pressure to a manageable level, and thus β phase processing is necessary rather than the more desirable α+β phase processing. Note that α+β extrusion exhibited higher strength and ductility values compared to those of β extrusion but lower strength values than those of flowformed tubes.

Table II. Tensile Properties of Ti-6Al-4V Tubular Materials

Material	UTS (MPa)	YS (MPa)	Elongation (%)	RA (%)
α+β Extruded Preform	981	909	17.3	38.6
Flowformed-1	1,227	1,163	12.7	41.0
Flowformed-2	1,309	1,210	11.0	32.0
β Extruded	965	862	13.3	25.7
α+β Extruded	1,043	988	14.0	36.7

Fatigue Properties: Blanks from the tubes were extracted in the axial direction and further machined into flat fatigue specimens and tested in accordance with ASTM E 466 [13]. Fatigue testing was performed at a frequency of 60 Hz under fully reversed stress condition, i.e., at a stress ratio, R, of −1.0. The fatigue data, provided in Table III, were analyzed using a linear regression approach and stress versus number of cycles (S-N) curves were generated using the procedures described by the ASTM E 468 and ASTM E 739 standards [14, 15]. The runout test data (i.e., >10^7 cycles) were not included in the regression analysis as recommended by the standards but were included in the S-N plots. The results are displayed in Figure 7. Fatigue properties of the Flowformed-1 and 2 materials are comparable, although there is more scattering in the Flowformed-1 data. Both materials show better fatigue performance than that of the β-extruded material (Figure 7a). Welding lowered the fatigue performance of the Flowformed-1 material, particularly at maximum applied stress levels below 450 MPa (Figure 7b). α+β extruded tube exhibited higher fatigue strength values compared to those of β extrusion as summarized in Table III.

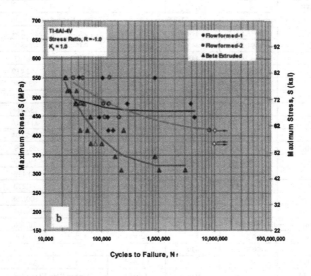

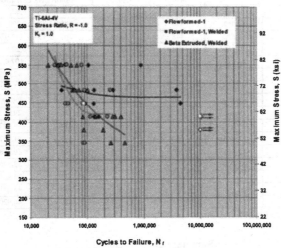

Figure 7. Fatigue properties of Ti-6Al-4V tubes: (a) Flowformed-1 and 2, and β-extruded materials, and (b) Flowformed-1 and β-extruded materials compared to the flowformed-welded materials.

Table III. Fatigue Properties of Ti-6Al-4V Tubes Produced by Flowforming and Extrusion

Max. Stress (MPa)	Cycles to Failure	Max. Stress (MPa)	Cycles to Failure	Max. Stress (MPa)	Cycles to Failure
Flowformed-1		**Flowformed-2**		**Beta Extruded, Welded**	
379	10,003,321*	414	8,226,820	345	263,234
379	10,091,640*	448	115,889	345	465,875
414	133,294	448	208,898	379	86,123
414	161,835	483	43,056	379	192,419
414	10,221,262*	483	110,567	414	83,959
448	92,807	483	139,734	414	140,917
448	130,975	550	32,213	414	282,768
448	4,509,510	550	47,379	414	311,916
483	35,299	552	141,976	414	395,504
483	132,963	**Beta Extruded**		483	55,179
483	291,682	310	223,292	483	59,390
483	3,872,958	310	1,050,748	483	60,459
550	105,913	310	3,052,626	483	71,270
550	898,496	345	173,618	483	100,758
552	40,871	345	876,520	550	20,707
Flowformed-1, Welded		345	903,163	550	30,766
345	85,229	379	64,833	550	52,390
345	89,576	379	78,118	550	66,169
414	114,027	379	101,899	552	34,639
414	227,726	414	42,050	**Alpha+Beta Extruded**	
414	10,057,921*	414	56,263	345	8,825,591
448	41,762	414	250,941	345	10,000,149*
448	46,550	448	63,405	362	144,076
448	83,609	448	64,285	362	6,203,273
448	86,382	483	35,162	379	107,439
483	60,054	483	37,170	379	62,099
483	90,199	483	49,787	414	77,395
483	257,956	517	25,689	414	69,200
550	26,766	517	28,257	448	29,993
550	31,585	517	36,883	448	49,837
550	32,945	550	23,147	483	51,777
550	81,046	550	24,038	483	30,298
552	44,046			517	33,546
				517	27,665
				550	43,051
				550	46,920

* Tests were stopped and called runouts.

Summary and Conclusions

Flowforming was utilized to produce seamless Ti-6Al-4V tubes along with several other processes including extrusion. Microstructure and mechanical properties of the resultant materials were studied. Microstructures of the representative materials were characterized utilizing optical microscopy, transmission electron microscopy and orientation imaging microscopy. The flowformed materials showed a very fine microstructure and severe deformation with a "fiber-like" microstructure consisting of severely elongated α and intergranular β. TEM study on a flowformed sample further verified very fine grains in both equiaxed and elongated morphologies with severe deformation through observed high dislocation densities. Tensile properties of the flowformed materials were impressively higher than those of β extruded material. A similar trend was observed in fatigue properties. In addition to producing superior mechanical properties, flowforming process provides tubes that require little or no machining. In summary, flowforming produces high tolerance, thin-wall, net-shape Ti-6Al-4V tubular products that are not possible to produce by competing processes such as extrusion, thus providing significant economic advantages.

Acknowledgement

This work was conducted by the Navy Metalworking Center, operated by Concurrent Technologies Corporation under Contract No. N00014-00-C-0544 to the Office of Naval Research as part of the U.S. Navy Manufacturing Technology Program. Approved for public release, distribution is unlimited.

References

1. www.flowformingplus.com, *Flowforming Overview*; *Understanding flowforming*, PMF Industries, Williamsport, PA, web-site accessed on October 10, 2006.
2. http://www.flowform.com/index.php, Dynamic Flowform, Billerica, MA, web-site accessed on October 10, 2006.
3. Mehmet N. Gungor, Ibrahim Ucok, Lawrence Kramer, Hao Dong, Nicholas Martin, Troy Tack, "Microstructure and Mechanical Properties of Highly Deformed Ti-6Al-4V," *The Langdon Symposium: Flow and Forming of Crystalline Materials,* eds. Y.T. Zhu, Z. Horita, K. Xia, P.B. Berbon, S.V. Raj, A.H. Chokshi, G. Kostorz, (Elsevier, 2005), 369-374.
4. Ibrahim Ucok, Lawrence Kramer, Mehmet Gungor, Philip Wolfe, Troy Tack, "Effect of Welding on Microstructure and Tensile properties of Flowformed Ti-6Al-4V Tube," *The Langdon Symposium: Flow and Forming of Crystalline Materials,* eds. Y.T. Zhu, Z. Horita, K. Xia, P.B. Berbon, S.V. Raj, A.H. Chokshi, G. Kostorz, (Elsevier, 2005), 160-164.
5. Kevin. Klug, Ibrahim Ucok, Mehmet Gungor, Mustafa Guclu, Lawrence Kramer, Troy Tack, Laurentiu Nastac, Nicholas Martin and Hao Dong, "Near-Net-Shape Manufacturing of Affordable Titanium Components for M777 Lightweight Howitzer: Research Summary," *JOM,* 56, No. 11, (2004), 35-41.
6. P. Finden, Proc. Sixth World Conference on Titanium, Part III, France, (1988), 1251-1256.
7. A. Wisbey et al., *J. Materials Science*, 27 (1992), 3925-3933.
8. J. Coyne in "The Science, Technology and Application of Titanium," Ed. R. Jaffee et. al., Pergamon Press, (1970), 97-110.
9. M. Donachie, Jr., Titanium: A Technical Guide, ASM International, OH, (2000), 31.
10. R. Miller, http://www.nsgrouponline.net/products/industry_news.asp, accessed October 12, 2006.
11. E. Ceretti et al., V AITEM 2001 Conference, Bari, Italy, Sept (2001), September 18-20.
12. ASTM E 8, Standard Test Methods for Tension Testing of Metallic Materials, ASTM, W. Conshohocken, PA, 2001, pp. 56-76.
13. ASTM E 466, Standard Practice for Conducting Force Controlled Constant Amplitude Axial Fatigue Tests of Metallic Materials, ASTM, W. Conshohocken, PA, 2000, pp. 493-497.
14. ASTM E 468, Standard Practice for Presentation of Constant Amplitude Fatigue Test Results for Metallic Materials, ASTM, W. Conshohocken, PA, 2000, pp. 508-513.
15. ASTM E 739, Standard Practice for Statistical Analysis of Linear or Linearized Stress-Life (S-N) and Strain-Life Fatigue Data, ASTM, W. Conshohocken, PA, 2000, pp. 631-637.
16. Same as reference 8, pp. 110-111.

Innovations in Titanium Technology

Novel Materials and Processes I

DEVELOPMENT OF THE FFC CAMBRIDGE PROCESS FOR THE PRODUCTION OF TITANIUM AND ITS ALLOYS

Richard Dashwood[1]; Martin Jackson[1]; Kevin Dring[2]
Kartik Rao[1]; Rohit Bhagat[1]; Douglas Inman[1]

[1]Imperial College London, London SW7 2BP, UK
[2]Norsk Titanium AS, Bankplassen 1a, N-0151 Oslo Norway

Keywords: FFC Cambridge Process, Electrochemistry, Titanium, Titanium Alloys

Abstract

The FFC Cambridge process involves the electrodeoxidation of TiO_2 in a molten $CaCl_2$ electrolyte to yield titanium metal. This process has attracted significant interest as a potential alternative to the Kroll process for the primary production of titanium or, when a mixed oxide precursor is used, as a low-cost method for producing alloy powder. For the last five years researchers at Imperial College London have been conducting fundamental studies to ascertain the electrochemical mechanisms involved in the process as well as evaluating the potential of the process to produce titanium alloys. This paper will demonstrate how electrochemical techniques such as linear sweep voltammetry have been used to elucidate the key thermodynamic aspects of the process. The paper will also present the results obtained from the production of titanium alloys highlighting the capability of the process to produce novel and complex alloys.

Introduction

The last ten years has seen the emergence of a number of novel reduction processes that have the potential to be low cost competitors to the established Kroll process [1]. However one process that has attracted arguably the most attention is the FFC Cambridge process. In September 2000, Fray Farthing and Chen (FFC), from Cambridge published a paper entitled "Direct electrochemical production of titanium dioxide to titanium in molten calcium chloride" in Nature and as a result the FFC Cambridge process emerged onto the world stage [2]. Since then interest from research institutions, funding bodies and commercial organizations has led to a flurry of activity in this area. The process involves the progressive reduction and deoxidation of a TiO_2 cathode in a calcium chloride melt. At the cathode the titanium oxide is reduced to titanium and the oxide ions dissolve into the calcium chloride. These oxide ions migrate to a carbon anode where they form carbon dioxide and carbon monoxide. Mixed oxides can be reduced in a similar manner to produce alloys [3]. Numerous research institutions have strived to elucidate the fundamental mechanisms of the process whilst a number of commercial organizations have explored the commercial potential of the process [4-6]. Imperial College London have been actively involved in research based on the FFC Cambridge process since September 2001 when DARPA/ONR commissioned a collaborative (with Cambridge University and QinetiQ) research program to evaluate the process. Since then work has progressed to incorporate mechanistic studies as well as investigating the reduction of mixed oxides to form novel and complex alloys. This paper will attempt to present the highlights of this research.

49

Experimental Procedure

The reduction cell (figure 1) consists of a programmable vertical tube furnace housing an Inconel® reaction vessel with a water cooled top plate. A 89 mm diameter OD, 79 mm diameter ID, 200 high Grade 2 commercial purity crucible was used to contain the molten calcium chloride. Prior to electrolysis the as-received anhydrous calcium chloride (Fluka) was subjected to a careful heating regimen in order to remove any physically adsorbed moisture. This involved heating the salt to 400°C and holding for at least 5 hours prior to a raising the temperature of the salt to the reaction temperature at 4°C.min^{-1}. During heating and subsequent electrolysis the salt was contained under argon which was dried by passing through silica gel and molecular sieves and oxygen gettered by passing over heated titanium turnings.

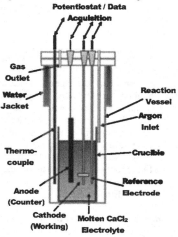

Figure 1. Schematic of the reduction cell apparatus.

Prior to electroanalytical experiments, pre-electrolysis (typically 2850 mV for 90 minutes) was performed using a graphite anode (10 mm diameter Tokai Carbon EC4) and titanium (3 mm diameter CP Ti rod) cathode to remove any electroactive impurities and residual moisture. Linear sweep voltammetry experiments were performed using a computer-controlled potentiostat (Princeton Applied Research EG&G Model 263A). A Ni^{2+}/Ni reference electrode was used for all voltametry work and consisted of a nickel wire immersed in a CaCl$_2$ containing 5 x 10^{-3} molal Ni^{2+} in a Pythagoras sheath [7]. For all experiments the counter electrode consisted of 10 mm diameter rods of Tokai Carbon EC4.

Material was analysed before and after reduction using a field emission gun scanning electron microscope (JEOL-FEGSEM) fitted with X-ray energy dispersive spectroscopy (X-EDS) capabilities operated at 20.0 keV. X-ray diffraction (XRD) patterns were obtained with a Philips PW 1710 diffractometer using α-Cu radiation and a Ni filter. Phase identification was conducted using the indexing cards published by the International Centre for Diffraction Data.

For the voltametry experiments the working electrode consisted of a continuous TiO$_2$ film (~10 µm thick) produced on a Grade 2 CP titanium rods by heating in air at 700°C for 36 h.

For the production of bulk titanium alloys the working electrodes consisted of 2 mm thick, 13 mm diameter porous (~ 50-60 % open porosity) disks. The disks were manufactured from reagent grade oxide powders from Alfa Aesar which were mixed with a small amount of distilled water and ground with a mortar and pestle for 300 seconds. The result mixture was uniaxially compacted in a 15 mm diameter die at 100 MPa and drilled to accept the cathodic current collector. The compacts were then placed in an alumina firing trough and sintered in air at 1100°C for 2 hours. The preforms were lowered into the electrolyte and reduced under applied voltages of 1500 mV for an initial period of 1 to 2 hours after which the voltage was increased to between 3000 and 3200 mV for the remaining duration of the 24 hour reduction time. Following reduction, the samples were raised from the molten salt into the upper regions of the reaction vessel and the reactor was slowly cooled to room temperature.

Results and Discussion

Mechanistic Studies

In order to elucidate the reduction mechanisms associated with the FFC Cambridge process a series of linear sweep voltagrams were performed at a range of temperatures (800-1100°C) and sweep rates (10-500 mV.s^{-1}). A typical voltagram obtained for a thin TiO$_2$ film (solid line) is given in figure 2. The sweep begins at the open circuit potential (in this case ~-250 mV), the cathode (TiO$_2$) is then polarized at a constant rate to more negative potentials and the resulting current recorded. Any current flow is indicative of electrochemical activity and any peak represents an electrochemical reaction.

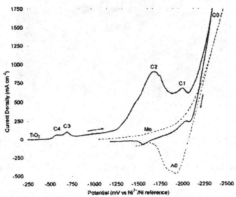

Figure 2. Linear sweep voltagrams conducted at 50 mV s^{-1} of TiO$_2$ and inert (Mo) working electrodes for cathodic-going scans in CaCl$_2$ at 800°C.

There are 5 clear peaks observed and, with the exception of C0, which represents the decomposition of the CaCl$_2$ to form Ca, the electrochemical reactions associated with the peaks were identified by removing the cathodes before and after each peak. The phases present were identified by a combination of XRD and SEM/X-EDS.

XRD conducted on material removed at C4 showed that a combination of Ti$_3$O$_5$, Ti$_2$O$_3$ and CaTiO$_3$ were present. The electrochemical process occurring at C4 was ascertained to be the reduction of TiO$_2$ to Ti$_3$O$_5$ (reaction 1) and it is believed that the non-stoichiometric Ti$_3$O$_5$ partially disproportionated to Ti$_2$O$_3$ and TiO$_2$ (reaction 2).

$$3TiO_2 + 2e^- \rightarrow Ti_3O_5 + O^{2-} \tag{1}$$

51

$$Ti_3O_5 \rightarrow Ti_2O_3 + TiO_2 \tag{2}$$

At C3 XRD detected no Ti_3O_5 but significant quantities of $CaTiO_3$ and Ti_2O_3 indicating that the electrochemical process associated with the C3 peak was the conversion of Ti_3O_5 to Ti_2O_3 (reaction 3). It is proposed that the $CaTiO_3$ is formed via a precipitation reaction involving the unreacted TiO_2, the Ca^{2+} from the electrolyte and the product oxide from reactions 1 and 3 (reaction 4). The presence of $CaTiO_3$ was visually confirmed by the blocky microstructure observed in a secondary electron image of the electrode surface (figure 3). The $CaTiO_3$ effectively forms a diffusion barrier preventing further significant electrochemical activity until the cathode is polarized to sufficiently negative potentials for the deoxidation of the $CaTiO_3$ phase. This accounts for the low currents between the end of C3 and the start of C2.

$$2Ti_3O_5 + 2e^- \rightarrow 3Ti_2O_3 + O^{2-} \tag{3}$$

$$TiO_2 + Ca^{2+} + O^{2-} \rightarrow CaTiO_3 \tag{4}$$

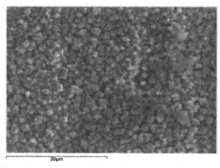

Figure 3. Secondary electron micrograph of cathode surface when removed at C3.

XRD on material removed at C2 identified the presence of TiO and Ti_2O. The latter is an ordered solid solution that is formed from cooling a disorded solid solution of titanium metal and oxygen. Clearly any $CaTiO_3$ and Ti_2O_3 formed at C3 was reduced to TiO via reactions 5 and 6.

$$CaTiO_3 + 2e^- \rightarrow Ca^{2+} + TiO + 2O^{2-} \tag{5}$$

$$Ti_2O_3 + 2e^- \rightarrow 2TiO + O^{2-} \tag{6}$$

At C1 the only phase identified via XRD is α-Ti indicating that reduction to metal has been completed via reaction 7.

$$TiO + 2e^- \rightarrow Ti + O^{2-} \tag{7}$$

This is not the end of the process as metallic Ti has a significant solubility for oxygen and this needs to be removed via reaction 8.

$$[O]_{Ti} + 2e^- \rightarrow O^{2-} \tag{8}$$

The gradual removal of oxygen from metallic titanium is demonstrated in figure 4 where two XRD traces obtained from material removed after a 30 second hold at potentials associated with peak C1 and C0 are shown. The shift in the peaks is a direct result of the removal of oxygen from the hexagonal close packed lattice. The oxygen content in the titanium can be calculated from this peak shift to be 6 wt.% at C1 and less than 1 wt.% at C0 [7].

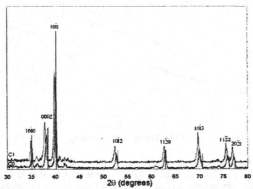

Figure 4. XRD patterns of TiO_2 working electrodes following a
30 second constant potential hold at the conclusion of linear
sweep voltammetry to either C1 or C0.

Broadly speaking the mechanism described above is in agreement with the findings of Schwandt and Fray [6] who took an alternative approach in determining the reaction pathway. The only discrepancy concerns the mechanism by which $CaTiO_3$ is formed. In the current work it is proposed that $CaTiO_3$ forms via a chemical reaction whereas Schwandt and Fray propose an electrochemical mechanism.

The thermodynamics of the Ca-Ti-O-Cl system are well characterized and following the method developed by Littlewood et al [8,9] it is possible to construct predominance diagrams for this system. Examples are given in figures 5 and 6 for temperatures of 1100 and 800°C respectively.

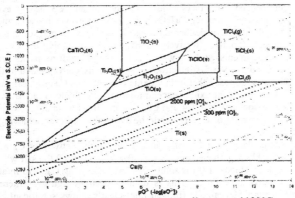

Figure 5. Ca-Ti-O-Cl predominance diagram at 1100°C

53

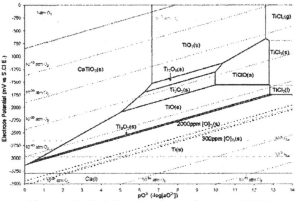

Figure 6. Ca-Ti-O-Cl predominance diagram at 800°C

The diagrams plot electrode potential on the y axis against the negative logarithm of the oxide activity in the salt (pO^{2-}) on the x axis and indicate which species should be present at different potential/oxide activity combinations. The bottom left hand corner of the graphs correspond to highly negatively polarized cathodes with a high O^{2-} concentration. What these diagrams show is the effect the oxide concentration in the salt has on the reduction process. Whilst significant attempts are made to ensure that the $CaCl_2$ is contaminant free, some oxide will always be present with a typical melt having a pO^{2-} of 3 at 900°C. Some key observations can be made from examination of these diagrams. As the oxide content in the salt increases or as the temperature decreases the deoxidation potentials are shifted to more negative values. This is important as it implies that to achieve the same level of deoxidation requires more negative potentials meaning that it becomes harder, or impractical (due to the decomposition limit of the salt) to produce low oxygen titanium. Furthermore at sufficiently high oxide activities $CaTiO_3$ is thermodynamically predicted to form via the chemical reaction of the TiO_2 working electrode and Ca^{2+} and O^{2-} ions in the melt. The effect of temperature and oxide activity have been confirmed by the results of linear sweep voltammetry [7]. The occurrence of $CaTiO_3$ is not so clear. Under standard conditions, where the pO^{2-} is in the order of 3, $CaTiO_3$ only forms once deoxidation has commenced (i.e. it coincides with the C4 and C3 peaks) even though the predominance diagrams would predict $CaTiO_3$ formation. However in experiments where the oxide content in the salt had been deliberately raised, to achieve a pO^{2-} of approximately 0.6, $CaTiO_3$ formed extremely rapidly, effectively passivating or consuming the TiO_2 film. In liner sweep voltammetry experiments working with this high oxide melt no peaks corresponding to C4 and C3 were observed and the first electrochemical activity corresponded to the deoxidation of $CaTiO_3$ (C2) [7]. It is proposed that in low oxide content salts $CaTiO_3$ formation is sluggish and only proceeds when the local oxide concentration at the cathode is raised as a result of the deoxidation reactions associated with C4 and C3. Clearly if a low enough oxide content could be achieved then the formation of $CaTiO_3$ could be avoided. However achieving this would in practice be difficult.

One further feature elucidated by these diagrams relates to the activity of Ca as a function of potential. Clearly at highly negative potentials (i.e. -3.1 V at 1100°C) the salt will decompose to produce unit activity Ca. However Ca is soluble in $CaCl_2$ allowing Ca to form at less than unit activities at less negative potentials. This is demonstrated in figure 2 where linear sweep voltammetry with an inert Mo cathode (broken line) clearly shows significant electrochemical activity at potentials less than that for the unit activity formation of Ca. On the predominance diagrams (figures 5 and 6) two lines have been drawn to represent the Ca activities of 10^{-3} and

10^{-6}. Considering the potentials where these lines appear the Ca activity associated with the polarization potentials for the initial stages of the reduction process will be so low that the kinetics of any calciothermic reduction could be so low as to be able to discount it as a possible mechanism. However in the later stages of reduction where the cathode reaches sufficiently negative potentials corresponding to relatively high Ca activities one cannot discount calciothermic reduction as being a potential mechanism. In fact Ono and Suzuki exploit calciothermic reduction in what is a similar process for the production of titanium [10].

Alloy Production

The FFC Cambridge process can also be exploited to produce alloys of titanium simply by blending a mixture of the required element oxides to form a cathode preform. In the work conducted to date at Imperial College work has focused on forming alloys with beta stabilizing elements and in particular isomorphous elements. The main reason for this is that, theoretically, beta stabilized alloys should reduce more readily as in the later stages of the process (where the oxygen is being removed from solid solution in metallic titanium) they will spend more time in the beta condition where oxygen diffusion is fastest.

In the current set of experiments attempts were made to prepare a Ti-10 wt.% Mo, Ti-10 wt.% W, and a Ti-10 wt. % V alloy. In the latter case, additions of iron and aluminum were made to produce the commercial composition of Ti-10V-2Fe-3Al. Examples of the microstructures obtained are given in figure 7. In all three cases very homogeneous fined grained material was obtained and the microstructures are what would be expected from a slowly cooled beta alloy with a hypo-eutectoid composition. In all three cases pro-eutectoid alpha decorates the prior beta grain boundaries with the prior beta grain interiors having transformed to a eutectoid alpha/beta structure. The target and achieved chemistry of the alloys (as measured by X-EDS) are given in table I and in all three cases the target chemistries were achieved to within ±0.5 wt.%.

In the case of the Mo and V containing alloys the degree of homogeneity achieved is not too surprising as the both V_2O_5 and MoO_2 form solid solutions with TiO_2 at the temperatures used for sintering. In both cases homogeneous $(Ti,Mo)O_2$ and $(Ti,V)O_2$ solid solutions were obtained prior to sintering. However, in the case of the Ti-10W, after sintering the tungsten was still present in the form of discrete ~5 μm diameter WO_3 particles [3]. Figure 8 shows a Ca-Ti-O-Cl predominance diagram with that for the Ca-W-O-Cl overlaid. It can be seen that reduction of WO_3 to WO_2 will occur before reduction of TiO_2 starts. At this point WO_2 will form a solid solution with TiO_2 to produce $(Ti,W)O_2$ allowing homogenization to take place. It is proposed that the diffusion of W in TiO_2 is significantly faster than in Ti giving rise to the degree of homogeneity that could not be achieved in the metallic state; W being a notoriously slow diffuser in Ti.

Significant densification of the preforms took place during reduction, which is a further manifestation of the enhanced diffusion associated with titanium in the beta phase. Finally, by producing the alloys in the solid state, the prior beta grain size achieved is several orders of magnitude less than that obtained through traditional melt processing.

55

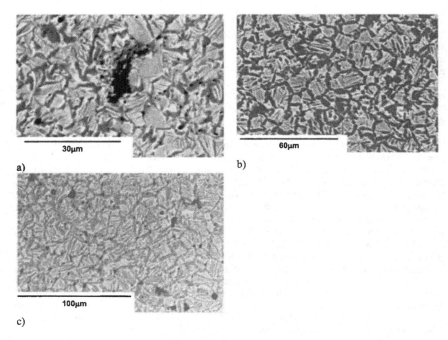

Figure 7. Backscattered electron images of reduced a) Ti-10Mo, b) Ti-10W and c) Ti-10V-2Fe-3Al

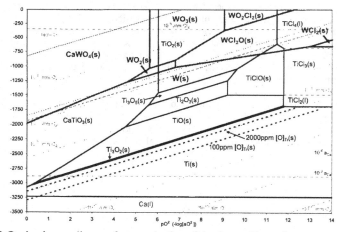

Figure 8. Predominance diagram for co-reduction of titanium oxides and tungsten oxides in molten $CaCl_2$ at 900°C.

Table I: Target and achieved chemistries (as measured by X-EDS) for alloys produced via the FFC Cambridge process.

Target (wt.%)	Achieved (wt.%)
Ti-10Mo	Ti-9.5Mo
Ti-10W	Ti-10W
Ti-10V-2Fe-3Al	Ti-9.6V-1.5Fe-3.1Al

Conclusions and Further Work

Liner sweep voltammetry has been used to identify the reduction sequence for the FFC Cambridge process as follows:-

$TiO_2 \rightarrow Ti_3O_5 \rightarrow Ti_2O_3 \rightarrow TiO \rightarrow Ti[O]$

At realistic oxide concentrations in the molten salt the formation of $CaTiO_3$ is unavoidable and will form a blocking layer which is stable up to highly negative potentials. Typically the $CaTiO_3$ will form chemically during the reduction to Ti_3O_5 and Ti_2O_3 due to a local increases in oxide concentration. In artificially high oxide concentration salts $CaTiO_3$ forms nearly instantaneously with the first electrochemical process being its reduction to TiO. Decreasing the temperature or increasing the oxide concentration of the salt has the effect of moving the reduction reactions to more negative potentials and increasing the stability of $CaTiO_3$ thus making the process less favourable and its ability to achieve low oxygen titanium harder. All the reduction steps take place at potentials more positive than that for the formation of unit activity Ca. However as Ca is soluble in $CaCl_2$ Ca can be formed, albeit at low activity, at potentials equivalent to those experienced in the FFC Cambridge process. Therefore the possibility of calciothermic reduction being involved in the later stages of the process cannot be ignored. Whilst the thermodynamics of the FFC Cambridge process are now generally well understood AC impedance techniques are currently being employed in an attempt to understand the kinetics of the process.

Alloys of titanium containing isomorphous beta stabilising elements have been successfully produced on the laboratory scale. The alloys are chemically and microstructurally homogenous as a result of the formation of oxide solid solutions during processing which can then homogenise relatively rapidly. The addition of the beta stabilising elements improve the efficiency of the process due to the fact that the titanium spends a longer period of time in the beta phase where diffusion rates are higher. Work is currently focussed towards reducing a sufficiently large batch of Ti-15Mo for solid state consolidation and subsequent mechanical property evaluation.

Acknowledgements

The authors are grateful to the Defence Advanced Research Projects Agency (DARPA) and the Office of Naval Research for funding and procuring the research contracts (DARPA N68171-01-C-9018 and DARPA N00014-04-1-0759). The authors appreciatively acknowledge support from the EPSRC (#EP/C536312).

References

1   E. H. Kraft, "Summary of emerging titanium cost reduction technologies", EHK Technologies, Report ORNL/Subcontract 4000023964, Oak Ridge National Laboratories, 2004.

2   G. Z. Chen, D. J. Fray, and T. W. Farthing, "Direct electrochemical reduction of titanium dioxide to titanium in molten calcium chloride", Nature, 2000, 407, 361-364.

3   K. Dring, R. Bhagat, M. Jackson and R. Dashwood, "Direct electrochemical production of Ti-10W alloys from mixed oxide precursors", Journal of Alloys and Compounds, 2006, 419, 103-109

4   I. Ratchev, S. Bliznyukov, R. Olivares, R. O. Watts, "Mechanism of electrolytic reduction of titanium dioxide in solid state in molten calcium chloride-based salts" Cost-Affordable Titanium: A Symposium Dedicated to Professor Harvey Flower as held at the 2004 TMS Annual Meeting; Charlotte, NC; USA; 14-18 Mar. 2004, 209-216

5   K. Dring, R. Dashwood and D. Inman: "Voltammetry of titanium dioxide in molten calcium chloride at 900°C" J. Electrochem. Soc., 2005, 152, (3), E104-E113

6   C. Schwandt, D. J. Fray, "Determination of the kinetic pathway in the electrochemical reduction of titanium dioxide in molten calcium chloride", Electrochimica Acta, 2005, 51, 66-76

7   K. Dring "Electrochemical reduction of titanium dioxide in molten calcium chloride" PhD Thesis, Imperial College London, 2005

8   R. Littlewood, "Diagrammatic representation of the thermodynamics of metal-fused chloride systems", J. Electrochem. Soc. 1962, 109, 525

9   K. Dring, R. Dashwood and D. Inman "Predominance diagrams for electrochemical reduction of titanium oxides in molten $CaCl_2$" J. Electrochem. Soc., 2005, 152, (10), D184-D190

10   K. Ono and R. O. Suzuki, "A new concept for producing Ti sponge: Calciothermic reduction", JOM, 2002, 54, 59-61

OPERATION OF ELECTROLYSIS CELLS FOR THE DIRECT PRODUCTION OF FERROTITANIUM FROM SOLID OXIDE PRECURSORS

Kevin Dring[1]; Odd-Arne Lorentsen[2]; Eirik Hagen[3]; Christian Rosenkilde[3]

[1]Norsk Titanium AS, Bankplassen 1a, N-0151 Oslo Norway
[2]Norsk Hydro - Oil and Energy, P.O. Box 2561, N-3908, Porsgrunn, Norway
[3]Norsk Hydro - Aluminum, P.O. Box 2561, N-3908, Porsgrunn, Norway

Keywords: Direct Electrochemical Production, Titanium, Molten Salt, Electrolysis

Abstract

Titanium and its alloys exhibit excellent mechanical properties, unrivalled corrosion resistance and outstanding biocompatibility; however, annual global titanium production is dwarfed by commodity metals. Since the 1950's several alternative processing routes to the Kroll method have been pursued, in vain; titanium oxides are extremely stable compounds that are bound with increasing tenacity to oxygen as the latter concentration decreases. Norsk Titanium, in collaboration with the research centers of Norsk Hydro in Porsgrunn, Norway, has extensively investigated the direct electrochemical electrolysis for the production of ferrotitanium from its constituent oxides. While immersed in a molten $CaCl_2$-based electrolyte, titanium-oxygen compounds were cathodically polarized against anodes of various compositions. The use of graphitic anodes resulted in the undesirable formation of molten carbonates, carbon scum on the surface of the electrolyte, and carbide contamination of the cathode product. As a consequence of the latter effects, a significant reduction in current efficiency was observed. Consequently, a variety of inert anode candidates have been evaluated, all of which feature the additional benefit of oxygen evolution as the anodic electrolysis product. The formation of titanium was found to proceed in a sequential manner, with the lower oxides of titanium, in conjunction with calcium titanate species, forming at intermediate reduction times. Formation of titanium was only observed to occur at the final stages of electrolysis, and the diffusion of oxygen out of the Ti-O solid solution was found to be rate determining.

Introduction

Norsk Titanium, in technical co-operation with Norsk Hydro, is developing a molten salt-based electrolytic process for the direct reduction of titanium dioxide to titanium metal. The research that will lead to the industrialization of this technology is being conducted at Norsk Hydro's Oil & Energy Research Centre in Porsgrunn, Norway and is currently conducting a multi-pronged program focusing on the design and development of electrolysis cells and its main constituents: the anode(s), cathode(s) and electrolyte. Other organizations are concurrently pursuing a similar solid-state approach to titanium production[1-3], whereby oxide precursors are progressively deoxygenated, and have reported mixed success in their commercialization efforts. This paper will discuss some of the challenges surrounding the development and optimization of the three major cell components.

Experimental Procedure

Molten salt reactors were constructed using a vertical anode and cathode arrangement in order to maximize the space-time efficiency of the cell. At the present scale, external heating elements surrounded an Inconel liner that housed an alumina or magnesia crucible, which was cast into place and subsequently sintered. A water-cooled lid supported a metal top plate that allowed the insertion of the anodes and cathodes, as well as reference electrodes, thermocouples and a salt sampling/addition tube. Inert gas purging could be supplied to the freeboard above the melt, as well as the cathode box during harvesting. On-line instrumentation recorded the electrode potentials, currents, cell voltage, temperature and power input. Periodic analysis of the melt oxide concentration was performed using acid titration.

The electrolytes studied in this system were initially pure calcium chloride with varying additions of calcium oxide. Additional experiments were conducted with the ternary $CaCl_2$-NaCl-CaO melts. The anode materials were either produced in-house from reagent grade powders or purchased from existing suppliers, such as Tokai Carbon for the anode blocks, which measured approximately 300 x 100 x 25 mm. The cathode precursors were extruded hollow shapes, either a honeycombed monolith or thick-walled (~1mm) tubes approximately 15-20 mm in width (figure 1). These were composed of rutile or ilmenite powders that had been mixed with organic binders and subsequently sintered under varying temperatures and times to obtain the desired porosity and ceramic microstructure. Numerous cathode geometries were assessed varying from wire binding of the preforms to a cathodic current collector to the use of a perforated cathode basket filled with the oxide precursors.

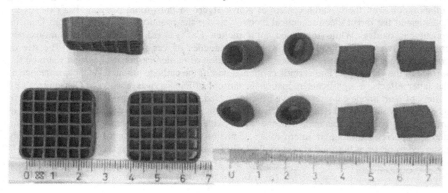

Figure 1. Ceramic precursors used in electrolysis cells (left: monolith, right: tubes).

Linear sweep voltammetry was conducted using a computer controlled PAR 273 potentiostat/galvanostat coupled to a CaO-buffered Ni/NiO reference electrode housed in a porous alumina sheath. Mo cavity electrodes were constructed from 99.5% pure foil (Alfa Aesar) and loaded with various reagent grade powders. The various materials were analysed before and after reduction using scanning electron microscopy with X-ray energy dispersive spectroscopy (X-EDS) capabilities and X-ray diffraction (XRD) techniques. Salt analysis was conducted using ICP, while the oxygen content of the titanium was evaluated using both XRD and Leco oxygen analyser.

Results and Discussion

Ferrotitanium Production

It was initially believed that the co-reduction of iron and titanium oxides to form ferrotitanium would permit the direct use of ilmenite ore as a ceramic precursor without the need to conduct the energy intensive removal of iron via processes such as titania slag production. A much cheaper feedstock could be used in this direct reduction process that would alleviate the current short supply of ferrotitanium, which is predominantly derived from titanium scrap. However, analysis of thermodynamic data for the Fe-Ti-O-Ca-Cl system (in the form of a predominance map with the activity of dissolved calcium on the y-axis and melt oxide activity on the x-axis) shows a complex mixture of phases and a high dependence of reaction pathway on the melt oxide content (figure 2)[4].

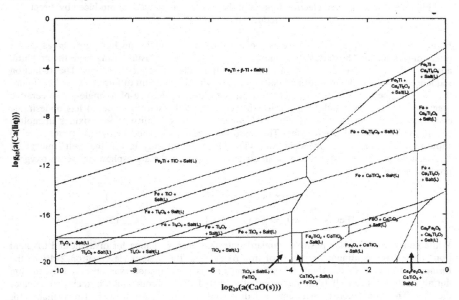

Figure 2. Predominance diagram for the system Fe-Ti-O-Ca-Cl at 900°C.

According to the predicted diagram, $FeTiO_3$ was only stable over a small potential and oxide activity window. Furthermore, the formation of calcium titanate and calcium ferrate species along with the dissolution of iron species into the salt phase needed to be considered during electrolysis. Alloys of iron and titanium were formed via direct reduction of ilmenite (figure 3), however, a stoichiometric 1:1 ratio of Fe:Ti was not the main phase formed. From the above figure, it is clear that the reduction of iron occurs at much lower dissolved calcium activities (less negative potentials) than the reduction of titanium-oxygen compounds, at a value of a_{Ca} of approximately 10^{-20} (a_{CaO} less than ~10^{-4}) increasing to 10^{-14} ($a_{CaO} = 1$). Consequently, the metallic iron can act as a conducting phase that increases the current distribution throughout the ceramic precursor. This is critical during the formation of calcium titanate compounds, which tend to passivate the cathode structure through both a mass-transfer, pore blocking mechanism and its intrinsic low electrical conductivity.

61

Figure 3. Backscattered electron images of the iron-titanium material produced by direct reduction of $FeTiO_3$.

In figures 3a & b, the white phase is the iron-rich metal, while the light grey areas depict titanium-rich metal. The darker grey phases in 3b are carbide deposits arising from the graphitic anodes, which will be discussed in further detail in the subsequent section. The reduction sequence shows that the iron formed does not alloy with the titanium to form FeTi or Fe_2Ti until the titanium metal has been sufficiently deoxidized, despite the use of a single-phase ceramic precursor. Unfortunately, the high iron oxide content of the starting material has undesirable consequences during electrolysis. Of chief concern is the solubility of iron oxides in molten calcium chloride-based electrolytes. The dissolution reactions listed below (reactions 1 & 2) show how the iron oxide enters the melt (1) and is subsequently lost at the melt/atmosphere interface when the cell is operated in an air environment or the atmosphere above an oxygen-producing anode.

$$FeO_x + xCaCl_2 \rightarrow FeCl_{2x} + xCaO \qquad (1)$$

$$2FeCl_{2x} + 1.5O_2 \rightarrow Fe_2O_3 + 2xCl_2 \qquad (2)$$

While equation 1 shows a positive standard state free energy change for divalent and trivalent iron species, the continuous removal of iron oxide from the system via equation 2 drives the former equilibrium to the right. Corroboration of this phenomenon was observed by fuming during electrolysis as well as the formation of reddish deposits throughout the freeboard space of the cell. The iron chloride that remained dissolved in the electrolyte was also available for reduction during the latter stages of the electrolysis when the cathode potential was sufficiently negative for electrodeposition. The "butterfly" structures seen in figure 3b are typical of the surface layer on the periphery of the ferrotitanium samples and illustrate the complexity of alloy production via the direct reduction of sintered mixtures of oxides. Additionally, the reduction of dissolved iron and titanium species at locations other than the cathode were observed and both Fe and Fe_2Ti were collected from the bottom of the crucible. This suggests that the soluble Fe and Ti species were reduced by dissolved calcium that had diffused away from the cathode and that the activity of dissolved calcium was steadily increasing in the bulk electrolyte as the electrolysis progressed. The solubility of the iron and titanium oxide components of the cathode precursors negatively affected mechanical integrity. The increase in cathode CaO content due to the formation of Ca-Fe-O and Ca-Ti-O compounds led to a swelling phenomenon that further decreased the mechanical strength of the cathode precursor. Weak precursors that were easily disintegrated were often retrieved from the cell when insufficient reduction had occurred and the dominant titanium oxide contained greater than 20 wt % oxygen.

Anode Development

The bulk of the patent literature associated with the direct electrochemical reduction of oxides to produce metals or alloys describes two types of anode materials: graphite and inert. While it is clear that graphitic anodes are useful in the characterization of the cathodic processes at the laboratory scale, several complicating factors demand that any industrial embodiment must use an inert anode material. While the use of carbon anodes to produce CO or CO_2 as the anodic reaction should provide a thermodynamic reduction in the overall cell voltage of approximately one volt versus pure oxygen evolution, a high overpotential partially offsets this economy in power consumption, while introducing several drawbacks pertaining to cell configuration and thermal balance, not to mention the production of greenhouse gases. The consumable nature of carbon anodes leads to a horizontal electrode arrangement in aluminum electrolysis, which is a considerably less efficient when considering the productivity per unit area of cell footprint. If a vertical arrangement is used to sidestep the problem, then thermal disruptions must be endured every time a "new" anode is inserted to replace a thinned, worn anode. Consequently, the interelectrode distance increases with time, giving a higher iR drop across the electrolyte and consuming more power disturbing the thermal balance of the cell since the position of the electrodes cannot easily be adjusted.

The use of carbon anodes at a larger scale was not as successful as at the gram scales reported in the initial discovery phases of these reduction processes. Severe carbon contamination was often observed on the exterior surfaces of the cathode precursors and cathode basket (figure 4). The mechanism of carbon contamination was via the formation of carbonate species in the vicinity of the anode (equations 3 & 4). Calcium carbonate would diffuse to the cathode, where the carbon would be reduced to form a carbide or carbon dust (equation 5).

Figure 4. Cathode basket before (left) and after (right) electrolysis using a graphite anode.

$$2O^{2-} + C \rightarrow CO_2 + 4e^- \qquad (3)$$

$$CO_2 + \underline{CaO} \rightarrow \underline{CaCO_3} \qquad (4)$$

$$\underline{CaCO_3} + 4e^- \rightarrow Ca^{2+} + 3O^{2-} + C \qquad (5)$$

Attempts to minimize the cathode current collector surface by suspending the precursors from the current collector with wire instead of using the cathode basket were unsuccessful. Additionally, very low current efficiencies were obtained when using graphitic anodes.

The difficulties in using graphite anodes may be attributed to two unavoidable consequences of the scale-up process: 1) lower quality carbon sources as the physical dimensions of the anode increase; and 2) lower electrolyte to feedstock volume ratios. The former may not necessarily be so, albeit at a cost penalty, since high-density carbon anodes may justify their greater expense. However, it is an intrinsic shortcoming of the carbon material itself, or rather, the circumstances under which it is operated, instead of any degradation attributable to poor microstructural or mechanical integrity of the anodes. This is not to say that mechanical degradation of the anode cannot lead to poor cell performance, as low density graphite anodes will tend to break apart with the result that the carbon floats to the electrolyte surface and short-circuits the cell. This problem is not severe in aluminum electrolysis as the horizontal anode occupies the upper portion of the cell. The restriction of decreased electrolyte volume relative to active cathode and anode areas/volumes is also one of economics since it is not feasible to operate a cell with large volumes of electrolyte supporting small electroactive regions. Consequently, saturation of the carbonate species will be much more rapid than in the instance where a large electrolyte volume is available to provide an effective "sink" and the times required to generate free carbon comparatively less.

A further factor in the formation of the carbonate species is the melt oxide content; with increasing melt CaO concentrations and decreasing temperatures, the solubility of carbonate increases – largely due to the solubility of dissolved CO_2. This further explains the discrepancies between the early experimental work, which was generally conducted on melts having very low CaO contents, and the scaled-up experiments, which employ melts with CaO contents greater than 1 wt %. The main reason for higher melt oxide contents is the need for an efficient production corresponding to high current densities, which increases the risk of chlorine formation at the anode during electrolysis. The limiting anodic current density for oxygen evolution is a function of both the bulk oxide concentration and the diffusion layer thickness. While the latter may be minimized by natural and/or forced convection in the cell, the limitations on the melt oxide activity cannot be ignored. At 800°C, each decade decrease in CaO activity (which is nearly equivalent to a decade change in concentration for melts containing less than 2 wt% CaO) results in approximately 100 mV positive shift in the oxygen evolution potential. As the standard state potentials for oxygen and chlorine evolution differ by approximately 600 mV at this temperature, the possibility of chlorine evolution may be easily realized at moderate anodic current densities.

From the perspective of oxygen evolution, a high melt CaO content is desirable, however, this is not so for the stability of the anode and cathode materials, which are generally oxide-based, since the solubility of the constituent oxides increases with increasing melt CaO concentration (figures 5a & b). Acidic oxides show a linear dependence of solubility versus melt CaO content and TiO_2 is one such oxide. Amphoteric oxides exhibit a minimum in solubility as a function of CaO concentration. At CaO contents less than this, the formation of metal chlorides is the dissolution mechanism, while at higher oxide contents, a metal oxychloride is favoured. This has significant consequences for the electrolyte CaO content since this mechanism shows how the anode constituents may dissolve and reach saturation in the electrolyte. Unfortunately, only a steady-state concentration for most impurities will be established since a complementary reaction will arise on the cathode material, especially where low oxygen titanium is present. The cathode basket and titanium metal will act as sinks for many impurities, leading to decreased physical properties and, more importantly, will cause the continuous dissolution of the anode material.

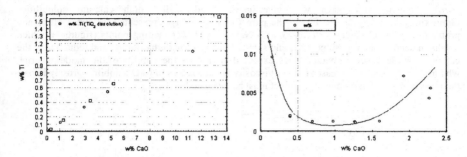

Figure 5. Solubility of acidic (left, a) and amphoteric (right, b) oxides in molten $CaCl_2$ at 900°C with varying CaO contents

The solubility of the oxide components of the cell as a function of melt CaO content is also related to the anode performance in the large electrolysis cells, regardless of the anode material chosen. The application of different current densities to the anode surface will create oxide concentration gradients of varying magnitudes across the Nernst diffusion layer (figure 6). At currents above the limiting current density for oxygen evolution, as shown by the bottom set of data points, some other charge transfer reaction must support the electrolysis current as mentioned earlier. Soluble species, M, in the electrolyte will precipitate as Ca-M-O out on the anode surface once the critical oxide concentration corresponding to the solubility product of the calcium-metal-oxygen compound is reached. Minimizing the formation of $CaTiO_3$ layers, which were observed on both carbon and oxygen-evolving anodes, remains an ongoing effort.

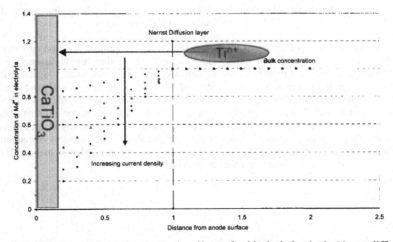

Figure 6. $CaTiO_3$ deposit formation due to the effects of oxide depletion in the Nernst diffusion layer during electrolysis.

65

Electrolyte Optimization

Initial experiments were conducted using CaCl$_2$-CaO melts in order to validate the findings of earlier research. Owing to the problems arising from high oxide solubilities in these melts, the ternary system with NaCl was investigated. Additionally, high calcium solubilities (approximately 4 at% at 900°C)[5] in the pure CaCl$_2$ electrolyte presented a current efficiency problem. CaCl$_2$-rich melts with 20-40 wt% NaCl offer lower melting points (figure 7), which results in decreased oxide solubilities and slower kinetics of both dissolution and reduction at the cathode[4]. While lower temperatures are expected to enhance the stability of the anode material, whether the associated increases in melt viscosity and decreases in CaO solubility grossly impair the mass transfer of the reduction product must be evaluated.

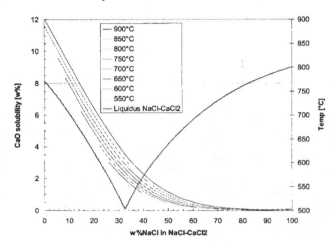

Figure 7. Phase diagram for the system NaCl-CaCl$_2$ showing the liquidus temperature and CaO solubility as a function of NaCl content.

Ongoing work is being performed to evaluate the nature of the cathode reaction pathway in the eutectic NaCl-CaCl$_2$ system with minor additions of CaO. Preliminary voltammetric studies show a change in the initial reduction process compared to that in molten CaCl$_2$ melts[6], with only one initial charge transfer process; the formation of Ti$_2$O$_3$ occurring in a one-step reduction from TiO$_2$ rather than proceeding through a Ti$_3$O$_5$ intermediary. To address the cathode kinetics at the lower temperatures studied in the ternary melts, the effect of varying additions of beta-stabilizing element oxides, such as V$_2$O$_5$ or MoO$_2$, on the reduction process is currently being evaluated. It is believed that the higher volume fraction of beta in these alloys will reach low-oxygen material more rapidly at a given temperature due to the approximately one order of magnitude higher diffusivity of oxygen in the beta phase versus the alpha phase.

Conclusions and Further Work

The direct electrochemical reduction of $FeTiO_3$ was observed to happen with the initial formation of metallic iron at the early stages of reduction, followed by the slower reduction of titanium oxides. Low-oxygen titanium was required before the alloying of iron and titanium would occur. The dissolution of the cathode constituents led to significant loss of iron to the melt and subsequently to the freeboard above the melt. Surface enrichment of iron resulted from electrodeposition of iron onto the cathode once a sufficiently negative potential was attained. Graphite anodes could not be successfully operated due to the low current efficiencies and carbon contamination on the cathode product. The development of an inert anode for this system is currently underway, although numerous technical challenges remain with respect to the material stability and optimum processing window.

References

1   R. O. Suzuki, K. Teranuma, and K. Ono, "Calciothermic reduction of Titanium Oxide and in-situ electrolysis in Molten $CaCl_2$", Metall. and Mater. Trans. B, 2003, 34, p287.
2   T. H. Okabe, T. Oda, and Y. Mitsuda, "Titanium Powder Production by Preform Reduction Process (PRP)", J. Alloys Comp., 2004, 364, p156.
3   Metalysis Press Release – October 2006, "Metalysis and BHP Billiton form joint venture company, Metalysis Titanium Inc", http://www.metalysis.com/news.htm
4   FactSage 5.4.1, Thermfact Ltd., Montreal, Canada, available from http://www.factsage.com/
5   K. M. Axler and G. L. DePoorter, "Solubility Studies of the Ca-CaO-$CaCl_2$ System", Mat. Sci. Forum, 1991, 73-75, p19.
6   K. Dring, R. Dashwood and D. Inman: "Voltammetry of titanium dioxide in molten calcium chloride at 900°C" J. Electrochem. Soc., 2005, 152, (3), pE104.

SYNTHESIS AND ENRICHMENT OF TITANIUM SUBCHLORIDES IN MOLTEN SALTS

Osamu Takeda[1] and Toru H. Okabe[2]

[1] Graduate School of Engineering, University of Tokyo;
7-3-1 Hongo, Bunkyo-ku; Tokyo 113-8656, Japan
(At present, Graduate School of Engineering, Tohoku University;
6-6-02 Aramaki Aza Aoba, Aobaku; Sendai 980-8579, Japan)
[2] Institute of Industrial Science, University of Tokyo;
4-6-1 Komaba, Meguro-ku; Tokyo 153-8505, Japan

Keywords: Titanium, Subchloride, Titanium Smelting, Molten Salts

Abstract

With the purpose of establishing a new (semi-)continuous/high-speed titanium production process based on the magnesiothermic reduction of titanium subchlorides ($TiCl_x$, x = 2, 3) referred to as the subhalide reduction process, a novel synthetic process for $TiCl_x$ by using molten salts as the reaction medium was investigated. Titanium tetrachloride ($TiCl_4$), supplied by a peristaltic pump at the rate of 0.13~0.72 g/min, was reacted with the feed titanium sponge immersed in molten magnesium chloride ($MgCl_2$) at 1273 K under an argon atmosphere. It was demonstrated that the efficiency of the formation of $TiCl_x$ was drastically improved by using molten salts as the reaction medium as compared with that of the synthesis by employing the direct reaction of $TiCl_4$ gas with solid titanium. Some results regarding the enrichment of $TiCl_x$ in molten salts are also shown.

Introduction

The rapid increase in the demand for titanium in recent years has resulted in a serious shortage of titanium in the market [1]. This is mainly because the productivity of the current titanium production process, called as the Kroll process [2–5], is low, and as such, an immediate increase in titanium production is difficult. In order to develop a new reduction process with high productivity, researches on the direct reduction processes of titanium dioxide (TiO_2) are actively being investigated worldwide [6–10]. These new reduction processes have the potential for producing low-cost titanium by simplifying the titanium production process. However, several technical problems such as impurity control need to be resolved prior to establishing a large-scale commercial process. Meanwhile, new titanium production processes based on chloride metallurgy are also being investigated [11–21] because chloride metallurgy has the essential advantage of producing high-purity titanium by carrying out the reduction process in an oxygen-free system.

Based on this background, the authors are developing a new (semi-)continuous/high-speed titanium production process based on the magnesiothermic reduction of titanium subchlorides—titanium dichloride ($TiCl_2$) and/or titanium trichloride ($TiCl_3$) [16–21]. This new titanium production process (cf. Figure 1), termed as the subhalide reduction process, involves three major steps: (1) synthesis of titanium subchlorides ($TiCl_x$, x = 2, 3), (2) enrichment of $TiCl_x$ when the subchlorides are produced in a magnesium chloride ($MgCl_2$) medium, and (3) production of titanium by the magnesiothermic reduction of $TiCl_x$.

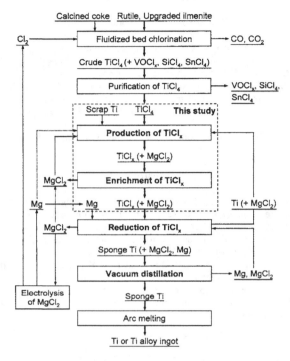

Figure 1. Flowchart of the new titanium production process based on the subhalide reduction process.

Titanium subchlorides have a fairly low vapor pressure, and they are stable as a condensed phase even at elevated temperatures. The advantages of utilizing subchlorides as feed materials for the reduction process are as follows. (a) A high-speed and (semi-)continuous process can be designed because the heat produced by the reduction of subchlorides is substantially lower than that produced by the reduction of titanium tetrachloride ($TiCl_4$) employed in the Kroll process. In addition, the density of the reaction field and the heat extraction rate from the reaction container increase drastically by employing the reduction process in the condensed phase. (b) The iron contamination of the titanium products from the reaction container can be avoided since metallic titanium can be utilized as the reactor material under $Ti/TiCl_2$ equilibrium. (c) The $MgCl_2$ reaction product can be easily removed by vacuum distillation, and high-purity titanium with a low oxygen content can be obtained. (d) The crushing of the massive sponge can be avoided when the size of the reactor is reduced, and the titanium obtained can be melted and cast into an ingot directly after the vacuum distillation.

Thus far, the authors have carried out fundamental research with the aim of establishing the subhalide reduction process that employs the magnesiothermic reduction of $TiCl_3$ or $TiCl_2$ in a titanium reaction container, and have successfully demonstrated the feasibility and advantages of the reduction process [16–20]. The authors have also developed a novel synthetic process for titanium subchlorides and have found that the efficiency of $TiCl_x$ formation was low when direct reaction of $TiCl_4$ gas with solid titanium was employed [21]. The authors are therefore developing a more efficient synthetic process for titanium subchlorides using molten salt [21]; some results of the development are introduced below. Furthermore, the experiments for the enrichment process of $TiCl_x$ in molten salt are also described.

Experimental

TiClₓ Synthesis in Molten Salt

The schematic illustrations of the experimental setup for $TiCl_x$ synthesis in molten salt are shown in Figures 2 and 3. Anhydrous $MgCl_2$ (97%, 100 g) dried under vacuum for more than 3 days was placed in a stainless-steel reaction container and heated to 1273 K in an argon atmosphere. $MgCl_2$ was expected to work as a medium that removes the solid $TiCl_2$ film formed on the surface of metallic titanium by dissolving and accumulating $TiCl_2$ in the medium. This is because the solubility of $TiCl_2$ in $MgCl_2$ at 1273 K is reported to be high (~83 mol%) [22]. After $MgCl_2$ had melted, the titanium sponge (99.7%, 6.3~19 g), which was placed in a stainless-steel basket, was immersed in the molten $MgCl_2$. Liquid $TiCl_4$ (99%, 26~76 g) was fed into this reaction container at the rate of 0.13~0.72 g/min by dropping it from the top of the chamber through a stainless-steel feed tube. It was expected that the $TiCl_4$ supplied in the form of a gas bubble at elevated temperatures reacted with the titanium sponge in the molten $MgCl_2$. After the experiment, the stainless-steel basket was removed from the molten salt. The molten salt in the reaction container and the residual titanium sponge in the basket were gradually cooled in a furnace and recovered in a nitrogen atmosphere at room temperature.

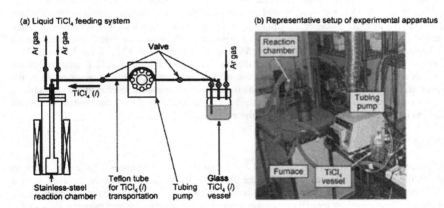

Figure 2. (a) Schematic illustration of a liquid $TiCl_4$ feeding system and (b) Photograph of the representative setup of the experimental apparatus.

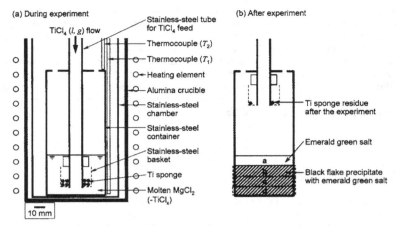

Figure 3.   Schematic illustration of the inner of the reaction container (a) during
and (b) after the experiment for $TiCl_x$ synthesis in molten $MgCl_2$ (Exps. W~Y).

Enrichment of $TiCl_x$ in Molten Salt

The vapor pressure of $TiCl_2$ at 1073 K (7.7×10^{-5} atm) is significantly lower than that at 1273 K (8.9×10^{-3} atm) [23]. Therefore, it was expected that $TiCl_2$ would be transferred from a higher temperature zone (with a high vapor pressure p_H) to a lower temperature zone (with a low vapor pressure p_L) to compensate for the vapor pressure of $TiCl_2$ in the entire system down to p_L when a temperature gradient exists in the $MgCl_2$–$TiCl_2$ molten salt (cf. Figure 4 (a)).

The $MgCl_2$–$TiCl_x$ salt (titanium concentration: 7.36~9.26 mass%, 15 g) obtained from the experiment for the $TiCl_x$ synthesis in molten salt was filled into a tubular stainless-steel capsule (SUS316, $\phi 9$ mm inner diameter, 200 mm length, cf. Figure 4 (b)), which was sealed by tungsten inert gas (TIG) welding. In the representative experiment, the entire capsule, which was placed in a horizontal furnace as shown in Figure 5 (a), was maintained at 1273 K for 3.6 ks (1.0 h). Subsequently, the temperature at one end of the capsule was decreased to 1012 K and maintained in this condition for 86.4 ks (24 h). After the experiment, the capsule was removed from the furnace and quenched in water (cf. Figure 5 (b)). In some experiments, the capsule was gradually cooled in the furnace after the experiment. The salt sample was then recovered from several locations in the capsule in a nitrogen atmosphere.

Analysis

The phases in the sample were identified by X-ray diffraction analysis (XRD). In order to prevent the samples from reacting with the moisture in the air, they were covered with a polyimide film in a glove box and then analyzed by XRD. The metallic element content in the sample was determined by inductively coupled plasma-atomic emission spectrometry (ICP-AES) after dissolving the sample in acid aqueous solutions. The chlorine content in the sample was determined by the potentiometric titration method. In this method, the sample was dissolved in an acid aqueous solution. A silver nitrate ($AgNO_3$) aqueous solution was then titrated into the

72

solution by monitoring the difference between the potential of a silver electrode and an Ag/AgCl electrode immersed in the solution.

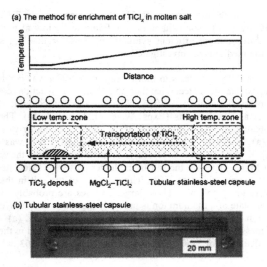

(a) The method for enrichment of TiCl$_x$ in molten salt

Figure 4. (a) Schematic illustration of the method for enrichment of TiCl$_x$ in molten salt and (b) Photograph of a tubular stainless-steel capsule before experiment.

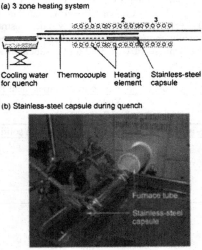

(a) 3 zone heating system

(b) Stainless-steel capsule during quench

Figure 5. (a) Schematic illustration of a 3 zone heating system and (b) Photograph of a tubular stainless-steel capsule during quench.

Results and Discussion

TiCl$_x$ Synthesis in Molten Salt

From the results obtained in the previous study [21], the efficiency of the TiCl$_x$ formation by the direct reaction of TiCl$_4$ gas with solid titanium was found to be low at temperatures between 1073~1273 K. Therefore, a more efficient synthetic process for titanium subchlorides using molten salt as a reaction medium was investigated.

After the experiment for TiCl$_x$ synthesis in molten salt, pure TiCl$_3$ was condensed at the lower temperature part of the stainless-steel chamber that held the reaction container. After the experiment, it was observed that the solidified salt in the reaction container had separated into two layers—the upper and lower parts (Exps. W~Y, cf. Figure 3 (b)). The concentration of titanium in the salt obtained from the upper part of the reaction container was 1.01~2.33 mass% (samples W (a), X (a), and Y (a)), whereas that from the lower part was 6.59~9.93 mass% (samples W (b)~(d), X (b)~(d), and Y (b)~(d)). From these results, the values of x in the composition formula MgCl$_2$–TiCl$_x$ of the salt in the lower part were calculated to be 2.04~2.61, which was close to the composition of TiCl$_2$. The x values of the salt in the upper part of the reaction container were greater than 4. These large x values are probably due to an analytical error, but the oxidation state of a titanium ion is probably high at 4. A one-layer solidified salt, in which the concentration of titanium was 12.2~18.9 mass% (samples Z (a)~(d)), was obtained when three times by mass of TiCl$_4$ and the titanium sponge were utilized in the experiment (Exp. Z). The values of x calculated for MgCl$_2$–TiCl$_x$ of the salt were 2.17~2.63, which was also close to 2.

The yield of TiCl$_x$ and the Ti consumption ratio in the experiment for TiCl$_x$ synthesis in molten salt are summarized in Table I. The representative results for the yield of TiCl$_x$ and the Ti consumption ratio when the direct reaction of TiCl$_4$ gas with solid titanium was employed for TiCl$_x$ synthesis are also shown in the table for reference. When MgCl$_2$ molten salt was employed as the reaction medium, the yield of TiCl$_x$ was 46~64%, and the consumption ratio of the titanium sponge feed was 80~93%. From these results, it was demonstrated that the efficiency of the TiCl$_x$ formation was drastically improved by using molten MgCl$_2$ and that MgCl$_2$ can be used as the reaction medium for the synthesis of TiCl$_x$ feed for the subchloride reduction process.

Table I.   Yield of TiCl$_x$ and Ti consumption ratio in the experiment for TiCl$_x$ synthesis [21]

Exp. No.	TiCl$_4$ feed rate, r / g·min^{-1}	Yield of TiCl$_x$, R_{TiCl_x} (%)	Ti consumption ratio, R_{Ti}' (%)
TiCl$_x$ synthesis by direct reaction of TiCl$_4$ gas with solid titanium			
A	0.64	26	43
B	0.17	32	44
C	0.17	23	42
D	0.12	35	45
TiCl$_x$ synthesis in molten salt			
W	0.56	64	80
X	0.16	46	93
Y	0.13	49	84
Z	0.65	60	75

Since the solubility of $TiCl_4$ in $MgCl_2$ was reported to be low (0.55 mol% at 1073 K) [24], it was expected that the $TiCl_4$ bubble injected into the molten $MgCl_2$ physically contacted with the metallic titanium to form $TiCl_2$, and that subsequently, the $TiCl_2$ formed dissolved into the molten $MgCl_2$. Meanwhile, the gaseous $TiCl_4$ could react with the $TiCl_2$ dissolved in the molten $MgCl_2$ to form $TiCl_3$ (cf. Eq. (1)), which in turn could react with the metallic titanium (cf. Eq. (2)). The detailed mechanism of the reaction of $TiCl_4$ with metallic titanium in molten salt is under investigation.

$$TiCl_2\ (s) + TiCl_4\ (g) \rightarrow 2\ TiCl_3\ (g) \tag{1}$$
$$\Delta H^{\circ} = +181\ kJ, \quad \Delta G^{\circ} = -28.4\ kJ\ \text{at}\ 1273\ K\ [23]$$

$$TiCl_3\ (g) + 1/2\ Ti\ (s) \rightarrow 3/2\ TiCl_2\ (s) \tag{2}$$
$$\Delta H^{\circ} = -212\ kJ, \quad \Delta G^{\circ} = +4.75\ kJ\ \text{at}\ 1273\ K\ [23]$$

<u>Enrichment of $TiCl_x$ in Molten Salt</u>

Since the concentration of $TiCl_x$ synthesized in the molten salt was low, it had to be enriched in order to improve the efficiency of its subsequent reduction by magnesium. For this reason, a fundamental experiment for investigating the feasibility of the enrichment process of $TiCl_x$ in molten salt was carried out.

In the representative experiment, the entire stainless-steel capsule containing $MgCl_2$–$TiCl_x$ salt was maintained at 1273 K for 3.6 ks (1 h). Subsequently, the temperature at one end of the capsule was decreased to 1012 K and maintained in this condition for 86.4 ks (24 h). After the experiment, the capsule was quenched in water. The analytical results for the concentration of titanium in the salt extracted from several locations in the capsule are shown in Figure 6 (a) along with the temperature profile before quenching. A black deposit was observed at the lower temperature part of the capsule, and its titanium concentration was 17 mass%, which was higher than the initial titanium concentration in the feed salt (8.9 mass%). The concentration of titanium in the salt around the black deposit was also high (13 mass%), although the titanium concentration in the salt at the higher temperature part remained almost unchanged from the initial value. When the capsule was gradually cooled in the furnace after the experiment, the concentration of titanium in the salt at the lower temperature part was high (15 mass%), while that at the higher temperature part was relatively low (6.6 mass%). This may be because the solidification of the salt started from the lower temperature part where $TiCl_2$ was preferentially solidified.

The mechanism of the enrichment of $TiCl_x$ is currently under investigation, but at this stage, its details are unclear. From these results, however, the feasibility of the enrichment process of $TiCl_x$ in molten salt by utilizing the temperature gradient was demonstrated, and this method can be utilized in the subhalide reduction process shown in Figure 1. Currently, technologies for enhancing the enrichment efficiency are also being developed.

75

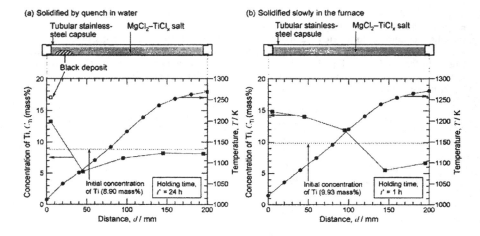

Figure 6.　Profile of the Ti concentration after the experiment for the enrichment of $TiCl_x$ in molten $MgCl_2$. Each sample was (a) solidified by quench in water [21] and (b) solidified slowly in the furnace after the experiment.

Conclusions

A fundamental study on new synthetic and enrichment processes for titanium subchlorides ($TiCl_x$, x = 2, 3) was carried out in order to establish a new (semi-)continuous/high-speed titanium production process based on the magnesiothermic reduction of titanium subchlorides (subhalide reduction process). It was demonstrated that the efficiency of $TiCl_x$ formation was drastically improved when molten salt ($MgCl_2$) was used as the reaction medium as compared with that of the synthesis by employing the direct reaction of $TiCl_4$ gas with solid titanium. Experiments for the enrichment process of $TiCl_x$ in molten salt were carried out, and the feasibility of the process was also demonstrated. The method for titanium subchloride production investigated in this study can be applied to the new high-speed titanium production process.

Acknowledgment

The authors are grateful to Professors Masafumi Maeda, Yoshitaka Mitsuda, and Kazuki Morita of the University of Tokyo and Professor Tetsuya Uda of Kyoto University for their valuable discussions throughout this project. The authors also thank Mr. Susumu Kosemura, Mr. Masanori Yamaguchi, and Mr. Yuichi Ono of Toho Titanium Co., Ltd. for their valuable discussions and for providing the samples. This work was financially supported by a Grant-in-Aid for Scientific Research (B) from the Ministry of Education, Culture, Sports, Science and Technology, Japan (MEXT, Project ID. #18360365). One of the authors, O. Takeda, is grateful for the financial support provided by a Grant for the 21st Century COE Program "Human-friendly Materials based on Chemistry" from MEXT and the financial support provided by Research Fellowships of the Japan Society for the Promotion of Science for Young Scientists.

References

1. S. Nakamura, *Industrial Rare Metals* (Arumu Publishing Co., Tokyo, Japan, 2004), 52–55.

2. W. Kroll, "The Production of Ductile Titanium", *Tr. Electrochem. Soc.*, 78 (1940) 35–47.

3. T. Ogasawara, "Progress of the Titanium Production Technology in Japan and Future Prospects of the Field", *Titanium Japan* (The Japan Titanium Society), 53 (2) (2005) 103–108.

4. A. Moriya and A. Kanai, "Titanium Sponge Production at Sumitomo Sitix Corporation", *Shigen-to-Sozai*, 109 (12) (1993) 1164–1169.

5. T. Fukuyama et al., "Production of Titanium Sponge and Ingot at Toho Titanium Co., Ltd.", *Shigen-to-Sozai*, 109 (12) (1993) 1157–1163.

6. Z. Chen, D.J. Fray, and T.W. Farthing, "Direct Electrochemical Reduction of Titanium Dioxide to Titanium in Molten Calcium Chloride", *Nature*, 407 (2000) 361–364.

7. K. Ono and R.O. Suzuki, "A New Concept for Producing Ti Sponge: Calciothermic Reduction", *JOM* (Journal of Metals), 54 February (2002) 59–61.

8. T. Abiko, I. Park, and T.H. Okabe, "Reduction of Titanium Oxide in Molten Salt Medium", *Proc. 10th World Conference on Titanium, Ti 2003* [Hamburg, Germany, July 13–18, 2003], (WILEY-VCH Verlag GmbH & Co. KGaA, Weinheim, Germany, 2004) 253–260.

9. T.H. Okabe, T. Oda, and Y. Mitsuda, "Titanium Powder Production by Preform Reduction Process (PRP)", *J. Alloys and Compd.*, 364 (2003) 156–163.

10. G.D. Rigby et al., "Polar^{TM} Titanium Development of the BHP Billiton Titanium Metal Production Process", *Presentation at 21st Annual ITA Conference, Titanium 2005* [Scottsdale, Arizona, USA, Sept. 25–27, 2005].

11. G.R.B. Elliott, "The Continuous Production of Titanium Powder Using Circulating Molten Salt", *JOM* (Journal of Metals), 50 September (1998) 48–49.

12. G. Crowley, "How to Extract Low-Cost Titanium", *Advanced Materials & Processes*, 161 November (2003) 25–27.

13. S. Takaya et al., "Titanium Production by Magnesiothermic Reduction of Molten $TiCl_2$–$MgCl_2$ Salts", *Abstracts of the Mining and Materials Processing Institute of Japan (MMIJ) 2005 Spring Meeting II* (MMIJ, Tokyo, Japan, 2005) 87–88.

14. A. Fuwa and S. Takaya, "Producing Titanium by Reducing $TiCl_2$-$MgCl_2$ Mixed Salt with Magnesium in the Molten State", *JOM* (Journal of Metals), 57 October (2005) 56–60.

15. M.V. Ginnata, "Titanium Electrowinning", *Proc. 10th World Conference on Titanium, Ti*

2003 [Hamburg, July 13–18, 2003], (WILEY-VCH Verlag GmbH & Co. KGaA, Weinheim, Germany, 2004) 237–244.

16. O. Takeda and T.H. Okabe, "New Titanium Production Process by Magnesiothermic Reduction of Titanium Subhalides", *Abstracts of EUCHEM 2004 Molten Salts Conference* [Poland, June 20–25, 2004], (Wydawnictwo Uniwersytetu Wroclawskiego Sp. z o. o., Wroclaw, Poland, 2004) 186.

17. O. Takeda and T.H. Okabe, "High Speed Reduction Process of Titanium Using Subhalide", *Abstracts of the Mining and Materials Processing Institute of Japan (MMIJ) 2004 Autumn Meeting II*, (MMIJ, Tokyo, Japan, 2004) 329–330.

18. O. Takeda and T.H. Okabe, "A New High Speed Titanium Production by Subhalide Reduction Process", *Proc. 2005 TMS Annual Meeting* [San Francisco, California, Feb. 13–17, 2005] (2005) 1139–1144.

19. O. Takeda and T.H. Okabe, "High-Speed Titanium Production by Magnesiothermic Reduction of Titanium Trichloride", *Materials Transactions*, 47 (4) (2006) 1145–1154.

20. O. Takeda and T.H. Okabe, "Fundamental Study on Magnesiothermic Reduction of Titanium Dichloride", *Metallurgical and Materials Transactions B*, 37B (2006) 823–830.

21. O. Takeda and T.H. Okabe, "Fundamental Study on Synthesis and Enrichment ofs Titanium Subchloride", *J. Alloys and Compd.*, submitted.

22. K. Komarek and P. Herasymenko, "Equilibria between Titanium Metal and Solutions of Titanium Dichloride in Fused Magnesium Chloride", *J. Electrochem. Soc.*, 105 (1958) 210–215.

23. I. Barin, *Thermochemical Data of Pure Substances*, (VCH Verlagsgesellschaft mbH, Weinheim, Germany, 1989).

24. M.V. Smirnov and V.S. Maksimov, "Solubility of Molten Titanium Tetrachloride in Molten Magnesium Chloride", *Electrochem. Mol. Sol. Electorolytes*, 7 (1969) 37–41.

PRODUCTION OF Al-Ti ALLOYS USING IONIC LUQUID ELECTROLYTES AT LOW TEMPERATURES

D. Pradhan[1] and R. G. Reddy[2]

[1]Graduate Student, [2]ACIPCO Professor and Head
[2]Associate Director of Center for Green Manufacturing
[1, 2]Department of Metallurgical and Materials Engineering
The University of Alabama, Tuscaloosa, AL 35487, USA

Keywords: Ionic liquid, Electrodeposition, Al-Ti alloy

Abstract

Application of novel low temperature ionic liquids for the extraction of metals was investigated. The present study is focused on production of Al-Ti alloys from $AlCl_3$-1-Methyl-3-butylimidazolium chloride (BmimCl)-$TiCl_4$ ionic melt at low temperature. The Al-Ti alloy was deposited on copper cathode at 1.5–3.0 V and 100 ± 3°C using the above melt at a molar ratio of 2:1:0.019 ($AlCl_3$: BmimCl: $TiCl_4$). Titanium sheet (>99.9 wt %) was used as the anode and would serve as the source of titanium apart from the $TiCl_4$ added to make the electrolyte. Morphology and composition of the produced Al-Ti alloy was investigated using SEM-EDS and phase analysis was carried out using XRD. The Al-Ti alloys containing about 15-35 wt % Ti were produced.

Introduction

High-temperature creep resistance of pure aluminum can be improved significantly by alloying with small amounts of copper, magnesium, silicon, manganese, titanium, vanadium, zirconium and other elements. Titanium is one of the transition element that improves the corrosion resistance of aluminum significantly [1]. Al-Ti alloys are extensively used for aerospace applications because of their light weight, high strength-to-weight ratio and high temperature oxidation resistance.

Various processing methods were reported for producing Al-Ti alloys. The electrodeposition of titanium from high temperature molten chloride/fluoride melt [2-8] was extensively investigated. However, the electrodeposition of Al-Ti alloy at room temperature or very low temperature was reported recently [9-11]. In recent years, Al-Ti alloy deposition from room temperature ionic liquids (RTIL) melts gaining interest because of the unique chemical and physical properties such as wide temperature range for the liquid phase, high thermal stability, negligible vapor pressure, low melting point and wide electrochemical window.

The electrochemical production of Ti and Al-Ti alloy is complicated due to the variable oxidation states of titanium. The electrochemical behavior of titanium in Lewis acidic chloroaluminate melt was investigated [12-13]. The electrochemistry of TiCl4 in strongly Lewis acidic at room temperature molten salt (AlCl3-EtMelmCl) was reported [14]. It was found that the reduction of tetravalent titanium occurs in three consecutive steps. Ti (IV) first reduced to Ti (III) and then subsequently reduced to Ti (II) and Ti (0). But when the concentration of Ti (III) i.e. β-TiCl3 exceed the solubility limit then Ti (III) was passivated on the electrode surface and also precipitated [10, 11]. This passive film blocks the electrodes and prevents oxidation and reduction. It was reported that at higher applied potential this film breaks down and produce Ti (IV).

Recently, Reddy et al. [15-18] investigated the electrorefining of aluminum alloy in 1-Methyl-3-butylimidazolium chloride (BmimCl) and also produced high-purity aluminum deposit from $AlCl_3$-BmimCl melt at 100°C. When BmimCl is present in molar excess over $AlCl_3$ i.e. [$AlCl_3$: BmimCl] <1.0, the ionic liquid becomes basic. When the molar ratio is 1.0 it is neutral melt and the melt contains only $AlCl_4^-$ anions. But when $AlCl_3$ is present in molar excess over BmimCl i.e. [$AlCl_3$: BmimCl] >1.0, the electrolyte composed of $AlCl_4^-$, $Al_2Cl_7^-$, $Al_3Cl_{10}^-$ and other higher order anions. Only $Al_2Cl_7^-$ is subsequently reduced to aluminum at the cathode. In this study, we report the electrodeposition of Al-Ti alloy from the acidic melt of $AlCl_3$-1-Methyl-3-butylimidazolium chloride (BmimCl)-$TiCl_4$ at 100 ± 3°C.

Experimental Procedure

The procedure for synthesizing 1-Methyl-3-butylimidazolium chloride (BmimCl) ionic liquid was described in detail elsewhere [17]. Anhydrous $AlCl_3$ (Alfa Aesar®, 99.985 %) was used without further purification. Chloroaluminate melts were prepared by mixing weighed quantities of $AlCl_3$ to the BmimCl ionic liquid into a pyrex electrochemical cell. All weighing, melt preparation and experiments were carried out in argon filled Labconco® glove box. $AlCl_3$ was added stepwise into the BmimCl in small amount because the reaction is highly exothermic, ensuring that the temperature of the melt should not rise above 100°C. Continuous stirring was done to dissolve $AlCl_3$ fully in the ionic liquid. After dissolution of $AlCl_3$ in the melt, a dark brown colored melt formed. Anhydrous $TiCl_4$ (Sigma-Aldrich®, 99.9%) was used as received. Precise quantity of liquid $TiCl_4$ was injected into the chloroaluminate melt by stainless steel syringe. Care was taken because $TiCl_4$ is highly hygroscopic and vaporizes quickly.

All experiments were conducted in a 40 ml Pyrex® glass beaker fitted with Teflon/Perspex cap. The schematic diagram of the experimental setup is shown in Fig. 1. Copper sheet (15 mm width, 0.2 mm thick), titanium Sheet (15 mm width, 0.2 mm thick) and Ti wire (0.5 mm diameter) were used as cathode, anode and reference electrode, respectively. The cathode area of deposition and anode working area were calculated from the portion of electrodes immersed into the melt. Reference electrode was used to measure the electrode potential of anode and cathode individually, using a multimeter (Keithley Instrument Inc®). The constant potential power source (Kepco®) supplied the required potential across the electrodes. The experimental setup has a provision for inserting thermometer into the melt to monitor temperature. Experiments were conducted in the argon filled glove box with pressure slightly higher than atmospheric pressure to ensure that there should not be any leakage of $TiCl_4$ vapor from the melt. The electrolyte was stirred at a constant speed using a magnetic stirrer and the temperature was controlled by a hot plate. The melt was kept for at least 15-20 minutes for temperature stabilization before starting the experiment. All experiments were carried out at 100 ± 3 °C for duration of 4 hours.

The morphological characterization and quantitative analysis of the electrodeposited Al-Ti alloy samples were carried out using Philips® XL30 scanning electron microscopy (SEM) with attached EDAX facility. The X-ray diffraction (XRD) analysis of the electrodeposited alloy was performed out using Philips® PW3830 X-ray diffractometer which uses monochromatic CuK_α ($\lambda = 1.5406$ Å).

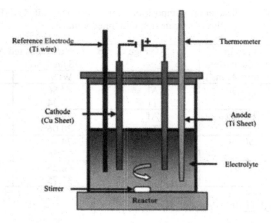

Fig 1. Schematic diagram of the experimental setup.

Results and Discussion

Electrodeposition of Al-Ti alloy

The bulk electrodeposition of Al-Ti alloy was carried out at constant applied potential using the melt containing $AlCl_3$, BmimCl and $TiCl_4$ with molar ratio of 2:1:0.019 ($TiCl_4$ concentration is 0.05m/l) at the temperature of $100 \pm 3°C$. The average compositions of the deposited Al-Ti alloy are reported in the table 1. The average composition was obtained from the EDS analysis of the deposit from four different regions on the sample. The above mentioned melt is highly acidic (molar ratio of $AlCl_3$: BmimCl = 2:1), and mainly consists of $Al_2Cl_7^-$ and other higher order ions. The $Al_2Cl_7^-$ ions traverse through the melt by diffusion, convection or migration and reduced at the cathode to produce aluminum deposit [18].

The cathodic aluminum deposition reaction is:

$$Al_2Cl_7^- + 3e^- \rightarrow Al \text{ (cathode)} + 7\ AlCl_4^- \qquad (1)$$

Hussey et al. [11] and Ali et al. [10] have reported that tetravalent or divalent titanium is present in the form of $[Ti\ (AlCl_4)\ _x]^{4-x}$. The reduction of tetravalent titanium occurs in three steps Ti (IV) $\rightarrow$Ti (III) $\rightarrow$Ti (II) $\rightarrow$Ti (0). We have observed that at high concentration of $TiCl_4$, a violet colored precipitate was formed at the bottom of melt and even on the electrode surface. This is due to the solid $TiCl_3$ which is sparingly soluble in the melt. Similar result was found by Hussey et al. [11], Freyland et al. [19] and Ali et al. [10]. Anodic dissolution of titanium was observed in all our experiments.

81

Table 1. The average composition of the electrodeposited Al-Ti
alloys as a function of applied voltage.

Applied potential (V)	Average cathodic current density (A. m^{-2})	Average Composition of deposited Al-Ti Alloy (EDS analysis)		
		Element	Wt %	Atom %
3	127.30	Al	82.66	89.43
		Ti	17.34	10.57
2.5	113.89	Al	79.94	87.60
		Ti	20.06	12.40
2	50.89	Al	74.44	83.78
		Ti	25.56	16.22
1.5	19.91	Al	62.25	74.53
		Ti	37.75	25.47

Characterization of Al-Ti alloy deposit

The morphology of Al-Ti alloy deposits and their composition analysis at four different applied potential are shown in Fig. 2 and Fig. 3, respectively. The SEM micrographs show that the deposits are nodular but not so dense or compact. The EDS analysis confirms the presence of Al and Ti on copper cathode. It was clearly seen from Fig. 2 and Table 1 that as the applied potential decreases from 3.0 to 1.5 V, the titanium content in the Al-Ti alloy increases from 17.34 to 37.75 wt% (10.57 to 25.47 at.%) and average nodule size or particle size decreases. But typically at higher applied potential the average nodule size should be smaller which conflicts our observation. This can be explained by considering nucleation phenomenon. At lower applied potential, especially at 1.5 V, titanium content in the alloy increases drastically which leads to increase the nucleation sites and produce finer nodules.

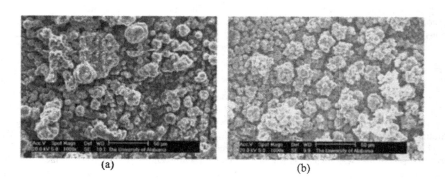

(a)   (b)

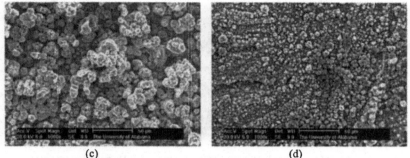

(c)   (d)

Fig. 2. SEM micrographs of Al-Ti electrodeposit at different applied potentials
(a) 3.0 V, (b) 2.5 V, (c) 2.0 V and (d) 1.5 V.

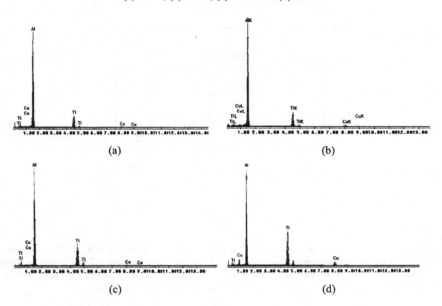

Fig. 3 EDS analysis of corresponding micrographs in Fig. 2 of Al-Ti electrodeposits
at different applied potential (a) 3.0 V, (b) 2.5 V, (c) 2.0 V and (d) 1.5 V.

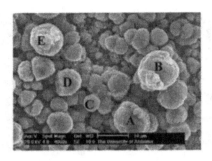

Fig. 4 SEM micrograph of Al-Ti electrodeposit at 3.0 V applied potential.
A, B, C, D, and E are different nodules where EDS analysis were carried out.

Table 2. The EDS analysis of nodules shown in Fig. 4.

Nodules		Composition of nodule (EDS analysis)	
		Wt %	Atom %
A	Al	86.58	91.97
	Ti	13.42	8.03
B	Al	86.64	92.01
	Ti	13.36	7.99
C	Al	84.70	90.77
	Ti	15.30	9.23
D	Al	87.19	92.36
	Ti	12.81	7.64
E	Al	81.01	88.34
	Ti	18.99	11.66

Further analysis on the composition of the deposited alloy was performed using EDS on different particles (Fig. 4) and are presented in Table 2. It can be seen that all the particles have approximately identical elemental compositions. The analysis confirms EDS area scan analysis shown in Fig 3 and moreover, aluminum or titanium particles were not found.

The X-Ray diffraction (XRD) patterns of the electrodeposited Al-Ti alloys are shown in Figure 5. The XRD analysis shows the diffraction peaks of Al and Cu. The diffraction pattern of copper appears due to the copper cathode. From the Al-Ti phase diagram and compositional analysis tabulated in Table 1, it can be seen that Ti should be present in the deposit as Al_3Ti. Hussey et al. [11] observed that Ti dissolve in Al and forms a disordered fcc at room temperature. They have identified the (100) and (110) superlattice reflections of fully ordered $(L1_2)-Al_3Ti$ deposited at higher temperature (150°C) from $AlCl_3$-NaCl melt.

In the present study, Al-Ti alloy deposited at a temperature of 100±3°C cannot be identified using XRD analysis. It may be due to the disordered structure of Al_3Ti that is prevailing at the temperatures used in the current study. Hence, individual diffraction peaks for titanium alloy (Al-Ti) cannot be differentiated from that of aluminum. However, further analysis using X-ray photoelectron spectroscopy (XPS) will confirm the composition of the electrodeposited Al-Ti alloy. XPS and TEM analysis of the electrodeposited Al-Ti alloys is ongoing, along with the studies on the effect of experimental variables such as temperature, electrolyte composition and stirring rate on composition and yield of the output.

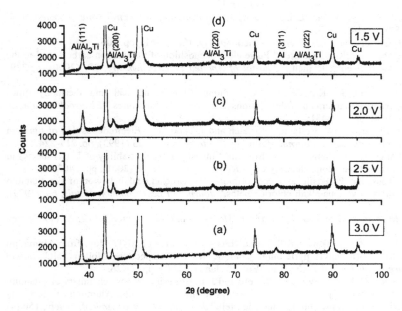

Fig. 5 XRD pattern of samples electrodeposited at different applied potential
(a) 3.0 V, (b) 2.5 V, (c) 2.0 V and (d) 1.5 V.

Conclusions

The production of aluminum-titanium alloy was successfully carried out on copper cathode from 2:1:0.019 molar ratio of $AlCl_3$-BmimCl-$TiCl_4$ melt at an applied potential ranging from 1.5 to 3.0 V at 100 ± 3°C. The product of Al-Ti alloy contained up to 37.75 wt% (25.47 at. %) of Ti. The alloy consisted of two phases Al and disordered Al_3Ti. Al-Ti alloy product at higher applied potential (3.0 V) becomes dark, non-uniform and no metallic luster was observed. But at lower applied potential (2.5V-2.0V) white deposit of Al-Ti alloy was obtained.

Acknowledgement

The authors gratefully acknowledge the financial support from ACIPCO and The University of Alabama.

References

1. G. S. Frankel, M. A. Russak, C. V. Jahnes, M. Mirzamaani, and V. A. Brusic, *Journal of Electrochemical Society*, 136, (1989), pp.1243.
2. D. M. Ferry, G. S. Picard and B. L. Tremillion, *Journal of Electrochemical Society*, 135, (1988), pp. 1443.
3. R. B. Head, *Journal of Electrochemical Society*, 108, (1961), pp. 806.
4. A. Robin and R. B. Ribeiro, " Pulse electrodeposition of titanium on carbon steel in the LiF-NaF-KF eutectic melt," *Journal of Applied Electrochemistry*, 30, (2000), pp.239-246,.
5. F. Lantelme, K. Kuroda and A. Barhoun, "Electrochemical and thermodynamics properties of titanium chloride solutions in various alkali mixtures," *Electrochimica Acta*, 44 (1998), pp. 421-431.
6. G. M. Haarberg, W. Rolland, Å. Sterten and J. Thonstad, "Electrodeposition of titanium from chloride melts," *Journal of Applied Electrochemistry*, 23 (1993), pp. 217-224.
7. L. Langrand, A. Chausse and R. Messina, "Investigations on stability of Ti (II) species in AlCl₃-dimethylsulfone electrolytes," *Electrochimica Acta*, 46 (2001), pp. 2407-2413.
8. E. Chassaing, F. Basile and G. Lorthioir. "Electrochemical behavior of the titanium chlorides in various alkali chloride baths," *Journal of Less-Common Metals*, 68, (1979), pp. 153-158.
9. G. M. Janowski and G. R. Stafford, *Metallurgical Transactions A*, 23A, (1992), pp. 2715.
10. M. R. Ali, A. Nishikata and T. Tsuru, "Electrodeposition of Al-Ti alloys from Aluminum Chloride-N-(n butyl) pyridinium Chloride Room Temperature Molten Salt," *Indian Journal of Chemical Technology*, Vol. 10, January 2003, pp. 21-26.
11. T. Tsuda, C. L. Hussey, G. R. Stafford and J. E. Bonevich "Electrochemistry of Titanium and the Electrodeposition of Al-Ti Alloys in the Lewis Acidic Aluminum Chloride-1-Ethyl-3-Methylimidazolium Chloride Melt," *Journal of Electrochemical Society*, 150 (4) (2003), pp. C234-C243.
12. A. Dent, K. Seddon and T. Welton, *Journal of Chemical Society, Chem Commum*, 4 (1990), pp. 315.
13. G. R. Stafford, "The Electrodeposition of Al₃Ti from Chloroaluminate Electrolytes," *Journal of Chemical Society*, 141, (1994), pp. 945.
14. R. T. Carlin, R. A. Osteryoung, J. S. Wilkes and J. Rovng, *Inorg. Chem.* 29, (1990), pp. 3003.
15. V. Kamavaram and R. G. Reddy "Thermal Stabilities and Viscosities of Low Temperature Aluminum Electrorefining Electrolysis: Di-alkylimidazolium Chloride Ionic Liquid," *Light Metals 2005, TMS*, 2005, pp. 501-505.
16. M. Zhang, V. Kamavaram and R. G. Reddy, "New Electrolytes for Aluminum Production-Ionic Liquids," *Journal of Metals*, (2003), pp. 54-57.
17. V. Kamavaram and R. G. Reddy, " Aluminum Electrolysis in Ionic Liquids at Low Temperature," *Metal Separation Technologies III*, R. E. Aune and M. Kekkonen (editors), Helsinki University of Technology, Espoo, Finland, (2004), pp. 143-151.
18. V. Kamavaram, D. Mantha, and R. G. Reddy, "Recycling of Aluminum Metal Matrix Composite using Ionic Liquids: Effect of Process Variables on Current Efficiency and Deposit Characteristics," *Electrochimica Acta*, 50, (2005), pp. 3286-3295.
19. I. Mukhopadhyay, C.L.R. Aravinda, D. Borissov, and W. Freyland, "Electrodeposition of Ti from TiCl₄ in the ionic liquid 1-methyl-3-butyl-imidazolium bis (trifluoro methyl sulfone) imide at room temperature: study on phase formation by in situ electrochemical scanning tunneling microscopy," *Electrochimica Acta*, 50, (2005), pp. 1275-1281.

Precipitation of Rutile (TiO$_2$) Nano-sized Particles via Forced Hydrolysis of a Titanium Tetrachloride (TiCl$_4$) Solution

Cécile Charbonneau, George P. Demopoulos

McGill University; 3610 University Street; Montreal, Quebec, H3A2B2, Canada

Keywords: Titanium tetrachloride, Forced hydrolysis, Rutile, Nano-sized particles

Abstract

This study investigates the production of titanium dioxide (TiO$_2$) nano-sized particles for advanced technological applications. During the preliminary phase of this project, the hydrolysis of titanium tetrachloride (TiCl$_4$) solution by heating under atmospheric pressure conditions has been examined. Monitoring of solution composition changes was accomplished by Inductively Coupled Plasma (ICP) spectroscopy and acid titration measurements. The precipitates were submitted to X-Ray Diffraction (XRD) analysis and Electron Microscopy observations. Agglomerates of crystalline nano-sized particles of rutile were formed. This paper will examine the influence of experimental parameters such as the temperature at which forced hydrolysis is carried out as well as the concentration of the TiCl$_4$ solution on the kinetics of the hydrolysis reaction and the characteristics of the solid products.

Introduction

Thanks to its superior opacifying and UV stabilizing properties titanium dioxide has been intensively produced and used in the industry since Second World War. This white pigment was first introduced into the plastic industry, and then later in the paint and paper industry for pigmentary purposes because of its additional good scattering properties, durability, chemical stability and lack of toxicity.

With the recent progression of nano-materials, new fields of applications were discovered in which titanium dioxide would be of a great interest. Not only titanium dioxide powders are recognized to be high quality white pigments in the particle size range 250-500nm, but it was found that nano-sized (<100nm) titania particles were endowed with remarkable UV absorption and photo-catalytic additional properties. The latter are being intensively investigated as keys in the fields of water treatment [1, 2], sterilization [3] and dye-sensitized solar cells [4, 5]. For instance, in the synthesis of dye-sensitized solar cells, the introduction of titanium dioxide potentially represents a much cheaper alternative to the high purity and so very costly materials utilized in current commercial solar cell technologies.

A diversity of new techniques has already been advanced for the synthesis of titania nano-powders with controlled properties (crystallographic form, size, shape...). For instance, thermal plasma synthesis [6, 7], chemical vapour deposition [8, 9] or spray hydrolysis [10, 11] were tried and tested to produce nano-sized crystalline powders and deposit thin films on substrates. However these techniques are very complex and energy-intensive, thus they are associated with high production cost. The sol-gel low temperature synthesis method has also been widely investigated for the preparation of titanium dioxide porous thin films and powders with large surface area [12, 13]. Nonetheless, sol-gel processes require the use of non-aqueous solvents; consequently they are potentially costly and not well suited for large scale production.

Direct precipitation of nano-sized TiO$_2$ out of aqueous TiCl$_4$ solutions without resorting to evaporation or use of organic solvents may provide a simpler, less costly and cleaner alternative to the methods mentioned above. Several investigations have been dealt with this approach in recent years. *Kim and Al* [14, 15] reported that crystalline either rutile or anatase (the two main naturally occurring crystalline forms of titanium dioxide), nano-sized particles could be produced by isothermally heating a diluted solution of TiCl$_4$. Brookite, another but less common crystalline form of TiO$_2$ was also reported by *Lee and Yang* [16] to be formed under specific hydrolysing conditions. The addition of reacting agents such as sulphate ions to the TiCl$_4$ solution and/or its neutralization with ammonia [17-20] were found to preferentially orient the hydrolysis reaction towards the synthesis of anatase nano-sized particles which are considered particularly suited for photo-catalytic applications.

The above review makes clear that production of TiO$_2$ powders with controlled properties by hydrolysis of TiCl$_4$ aqueous solution is of interest both for pigment as well as nano-material applications. Further work is needed to lead to optimization of such wet routes of synthesis, for example by avoiding the use of autoclaves and understanding the governing chemical phenomena. As a contribution towards this effort, we report preliminary results from a recently launched project which investigates the synthesis of nano TiO$_2$ phases by forced hydrolysis of TiCl$_4$ (aq).

Experimental

Titanium tetrachloride (99.9% TiCl$_4$, Fisher Scientific) was used as the main starting material without any further purification. In order to adjust the concentration of the solution, pure TiCl$_4$ was dropwise added to deionised water under magnetic stirring, until the desired stoichiometry was reached. The temperature of the system was maintained in the range 3-7°C by using an ice-water bath in order to prevent the violent exothermic reaction between liquid TiCl$_4$ and water. During the experiment, we could observe that as soon as TiCl$_4$ was coming into contact with the moisturized air, yellow cakes of orthotitanic acid Ti(OH)$_4$ and gaseous HCl were instantaneously forming, as quoted by *Kapias et Griffiths* [21]. Once entered in the aqueous phase, those yellow cakes dissolved back in a couple of seconds, generating a slight raise of the temperature of the solution. The latter would drop back as fast. The synthesized solution was used in a set of different experiments which are described below.

a) In the first set of experiments, we investigated the effect of temperature on the stability of TiCl$_4$ solutions of varied concentrations. 0.51M, 0.75M, 0.9M, 1.27M and 1.60M TiCl$_4$ solutions were gradually heated from 20°C to 95°C with a linear heating rate of 1°C/min. This was performed in a bath with constant magnetic stirring of the solutions. The starting point of precipitation was visually determined for each solution by observing the appearance of the first white cloud.

b) In the second set of experiments, only isothermal treatments were run. The effect of concentration and temperature on the kinetics of TiCl$_4$ hydrolysis was investigated: 0.51M, 0.9M and 1.60M TiCl$_4$ aqueous solution were isothermally treated at 80°C for 5 hours. These tests were repeated at 60°C and 70°C for the 0.9M TiCl$_4$ solution. Samples of 5ml approximate volume were regularly taken from each flask and directly submitted to a quench at 10°C in order to freeze the hydrolysis reaction. Half volume of each sample was filtered using 0.1μm pore size filters. The filtrates were then diluted in a concentrated solution of hydrochloric acid (3.24M HCl) in order to prevent any further precipitation. They were further submitted to ICP measurements so as to measure the concentration of Ti^{4+} remaining in solution. The other half of each sample, still containing precipitated TiO$_2$ particles was saved for particle size analysis. In the test where the 0.9M TiCl$_4$ solution was isothermally treated at 80°C, two larger samples were taken at 1h and 5h after the beginning of the experiment. The powders were filtered using 0.1μm pore size filters, washed twice in boiling deionised water and once in methanol and dried in a dessicator at ambient temperature (20°C) over night. The powders prepared in this manner were submitted to XRD and SEM analyses.

The concentration of Ti^{4+} in prepared or processed solutions was determined by inductively coupled plasma spectroscopy (ICP) measurements with a Thermo Jarrel Ash Trace Scan Machine. Standards of 50ppm, 500ppm and 1000 ppm of titanium and a blank (2% vol HNO_3) were used for the calibration. All standards were prepared from ICP grade standards of 1000 ppm. The powder particle size distribution (PSA) was determined using the HORIBA LA-920 Particle Size Analyzer. Deionised water was used as a medium. X-ray Powder Diffraction (XRD) analyses were performed on a Rigaku Rotaflex D-Max diffractometer equipped with a rotative anode, a copper target ($\lambda(CuK_{\alpha1}) = 1.5406$Å), a monochromator composed of a graphite crystal, and a scintillator detector. The diffractometer used 40kV and 20mA. Scanning took place between 2° and 100° (2θ) with a 0.1° step and an acquisition time of 3s per step. Scanning Electron Microscopy (SEM) and respectively Field Emission Gun Scanning Electron Microscopy (FEG-SEM) were used for the observation of the synthesized TiO_2 powders, on a JEOL-840 and respectively on a Hitachi S-4700 microscope. Prior to microscopic examination, the TiO_2 particles were deposited on a carbon film and coated with a thin layer of AuPd.

Results and Discussion

Experiment 1 – Effect of Temperature on Forced Hydrolysis

In this first experiment, the forced hydrolysis of $TiCl_4$ solutions as a function of increasing temperature and concentration was examined. As shown in Figure 1, at t=77min, the 0.51M and 0.75M solutions are very cloudy due to their advanced state of precipitation whereas the 0.94M solution only starts to show some cloudiness aspect and the 1.60M is still visibly transparent. This visual observation makes apparent that the stability of aqueous $TiCl_4$ under raising temperature or forced hydrolysis depends on the concentration of the solution. More precisely, the results reported in Figure 2 show that the higher the $TiCl_4$ concentration the higher the temperature of precipitation initiation and so the more stable the solution. A series of pH titrations using NaOH as a titrant and EDTA as an agent to prevent hydrolysis to occur, similarly to the method described by *Rolia and Dutrizac* [22], was also performed on a range of 0.01M to 2M $TiCl_4$ aqueous solutions. The results showed that the acidity of the solutions increases with the concentration of $TiCl_4$. The concentration of protons in the local neighbourhood of the Ti^{4+} cations might have a stabilizing effect and partially explain why more concentrated solutions require higher temperature to start hydrolysing.

According to literature [23], it is also possible that depending on the concentration of $TiCl_4$(aq), the nature of the species in solution may vary significantly and hence affect the chemical mechanisms involved in forced hydrolysis. In high concentration $TiCl_4$ solutions, complexes such as $[Ti_2Cl_9]^-$, $[Ti_2Cl_{10}]^{2-}$, $[TiCl_6]^{2-}$, $[Ti(OH)Cl_4]^-$, $[Ti(OH)_2Cl_4]^{2-}$, $[TiOCl_4]^{2-}$, $[TiOCl_3]^-$, have been detected whereas in low concentration $TiCl_4$ solutions complexes such as $[Ti(OH)_2]^{2-}$, $[Ti(OH)Cl]^{2-}$, $[Ti(OH)_6]^{2-}$, $[TiOCl_2]$, $[TiOCl]^-$, $[TiO]^{2-}$ are more likely to be found [24,25]. In high concentration $TiCl_4$ solutions, the presence of a higher number of large coordinated Cl^- anions probably makes it more difficult for water molecules to get coordinated to the Ti^{4+} cations and further deprotonated, preventing thus forced hydrolysis to occur. This assumption might partially explain why higher concentration solutions are more stable and require higher temperature to start hydrolysing. On the contrary, in low concentration $TiCl_4$ solutions, the coordinated Cl^- anions are more likely replaced by hydroxyl groups and water molecules which keep satisfied the octahedral environment of the metallic cations in solution and promote hydrolysis. The concentration, pH and nature of the complexes in solution seem to play an important role in the chemistry of forced hydrolysis.

More work needs to be done in order to fully understand the influence of each of these parameters on forced hydrolysis of TiCl$_4$ solutions and eventually to optimize the production of colloidal nano-sized TiO$_2$.

Figure 1. Experimental set up used in the first experiment (at t=77min, T=83°C), showing four conical flasks (a fifth one is hidden by flask a)) which are heated in a bath. Each of the flasks contains 50ml of aqueous TiCl$_4$ solution of initial concentration: a) 0.51M; b) 0.75M; c) 0.94M; d) 1.60M; e) 1.27M.

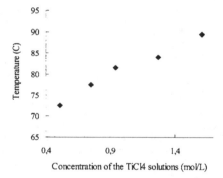

Figure 2. Diagram reporting the temperatures at which precipitation was visually determined to start in the five solutions of initial concentrations: 0.51M, 0.75M, 0.94M, 1.27M and 1.60M TiCl$_4$(aq).

Experiment 2- Precipitation Kinetics

In this second set of experiments, more emphasis was placed in examining the effect of initial TiCl$_4$(aq) concentration and temperature on the kinetics of hydrolysis in the case of isothermal transformations. For aqueous TiCl$_4$ solutions of 0.51M, 0.94M and 1.60M initial concentration, heated at 80°C for 5 hours, the converted fraction of TiCl$_4$ is reported as a function of time in Figure 3. The curve corresponding to the converted fraction of TiCl$_4$ in the 0.94M solution is sigmoidal in shape which is characteristic of isothermal phase transformations involving nucleation and growth [26]. In the first 30 minutes, the curve shows a pretty flat profile corresponding to the induction period, after which nucleation and growth of the particles occurs until the TiCl$_4$(aq) conversion is complete. In the case of the 0.51M solution the induction period is much shorter and the rate of particle growth is faster. At 15 minutes, we can notice that the converted fraction of TiCl$_4$ for the 0.51M solution is negative. This is due to the partial evaporation of water in the solution which induces an increase in the TiCl$_4$(aq) concentration. After 6 hours of isothermal treatment at 80°C, the 1.60M solution only has 20% of its TiCl$_4$(aq) content converted into TiO$_2$. These observations are in agreement with the results previously discussed for the first experiment: the higher the concentration of the solution the more stable is hence the slower the kinetics of forced hydrolysis. Figure 4 shows the effect of temperature on the kinetics of TiCl$_4$ forced hydrolysis. It appears that the conversion of TiCl$_4$ into TiO$_2$ is faster at higher temperature. The comparison of the curves corresponding to the 60°C and 80°C isothermal treatments makes clear that temperature mostly affects the nucleation period which is much longer at 60°C than at 80°C. Once nucleation is completed, the parallelism of the two curves suggests that the growth of the particles proceeds at similar rate in both cases.

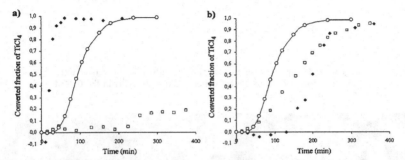

Figure 3. a) Evolution of the converted fraction of TiCl$_4$ into TiO$_2$ as a function of time during an isothermal treatment run at 80°C on 0.51M (♦), 0.94M (-O-) and 1.6M (□) TiCl$_4$(aq) solutions. b) Evolution of the converted fraction of TiCl$_4$ into TiO$_2$ as a function of time during isothermal treatments performed on a 0.94M TiCl$_4$(aq) solution at 60°C (♦), 70°C (□) and 80°C (-O-).

According to the mathematical model developed by Johnson, Melh, Avrami and Kolmogorov [26-28] to describe the kinetics of isothermal phase transformations, the converted fraction of reactant 'X' (TiCl$_4$(aq) in our case), can be expressed as a function of time as written in Equation 1.

$$X(t) = 1 - \exp[-(kt)^n] \qquad (1)$$

x(t): fraction of converted metal ion
k: kinetic parameter (cste)
t: time
n: Avrami exponent

The kinetic parameter, k, describes the rate of reaction, such as the nucleation and growth rates while the Avrami exponent, n, provides information on the dimensionality (one-, two- or three-dimensional) and the nature of the transformation. If the experimental data follow the JMAK model, the plot of ln[ln(1-X)-1] versus ln(t) gives a straight line with slope n and intercept n $ln(k)$. In order to characterize more precisely the kinetics of the mechanisms involved in the isothermal treatment of TiCl$_4$ solutions, the parameter n was calculated for the 0.94M solution submitted to an isothermal treatment at 80°C. The resultant plot is presented in Figure 4. The value calculated for *n* is 2.386, which according to the premises of Avrami's model is supposed to indicate a mix of spheroidal and nonspheroidal grown particles. This assumption matches with the SEM imaging analyses performed on TiO_2 precipitates which were synthesized from a 0.94M $TiCl_4$ solution after 3 hours isothermal treatment at 80°C: Figure 5(a) shows coarse and irregular TiO_2 particles of an average size of 2 to 6μm which result from the aggregation of smaller spherical particles of 1 to 2μm average diameter. Similar results were reported by *Sarraf-Mamoory et al* [29] in their study on the kinetics of copper particle nucleation and growth.

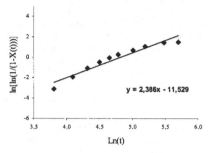

Figure 4. Diagram presenting a plot of ln[ln(1-X)-1] versus ln(t), where X corresponds to the converted fraction of $TiCl_4$ in the 0.94M $TiCl_4$ solution treated at 80°C.

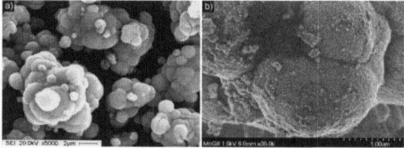

Figure 5. a) SEM picture of TiO_2 powders precipitated after 3 hours of isothermal treatment at 80°C applied to a 0.94M $TiCl_4$ solution. b) FE-SEM picture of TiO_2 powders precipitated after 1 hour of isothermal treatment at 80°C applied to a 0.94M $TiCl_4$ solution

Figure 6 gives the evolution of particle size distribution with time. The obtained particle size data can be understood with the help of the SEM and FE-SEM pictures respectively presented on Figure 5(a) and 5(b). The pictures suggest the occurrence of two aggregation stages following the nucleation stage. Although the particle size analyzer did not detect any particles smaller than 100nm, the picture presented in Figure 5(b) clearly shows that particles of an average size smaller than 50nm are first produced (Figure 6-15min) and aggregate to form bigger spherical particles of 1 to 2μm diameter. In addition a few non spherical particles of 100 to 200nm can also be seen on Figure 5(b). One hour later (Figure 6-1hour), the population of particles in the range size 100 to 200nm has disappeared whereas the population of a few microns size particles has increased and shifted towards a population of larger particles. This corresponds to a secondary stage of aggregation where spherical particles of 1 to 2μm diameter get aggregated to form coarse and non spherical particles of sizes ranging from 2 to

10μm. After this second stage of aggregation, the growth of the particles stabilizes and the particles size distribution becomes smoother and narrower (2-5 hours). The particles presented in Figure 5(a) are the outcome of the second stage of aggregation.

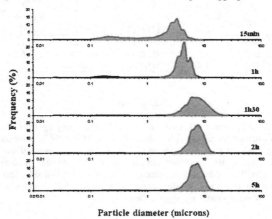

Figure 6. Evolution of the size distribution of TiO_2 particles precipitated out of a 0.94M $TiCl_4$ solution isothermally treated at 80°C for 5 hours.

The powders collected after 1hour and 5hour of isothermal treatment of a 0.94M $TiCl_4$ solution were submitted to XRD analysis. The crystalline phase which was determined to match the best with the corresponding patterns presented on Figure 7 was rutile.

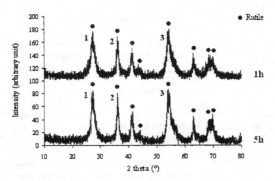

Figure 7. XRD patterns of TiO_2 powders obtained after 1 and 5 hours of isothermal treatment of a 0.94M $TiCl_4$ solution at 80°C

Both patterns look very similar, however comparison of the main peak features reveals a few differences: the relative intensities and heights of peaks #1 and #2 are smaller for the powders obtained after 1 hour treatment than for the powders obtained after 5 hours (See Table I). This can be interpreted as an increase of the crystalline fraction of the TiO_2 powders with the length of the isothermal treatment. The Full-Width Half-Maximum of the peaks is quite large due to the nano-size of the smaller formed unit crystallites. The typical size of these nano-crystallites was estimated to about 7nm by using the Scherrer formula [30].

93

Table I. Position, relative intensity and FWHM of three main peaks of the XRD patterns presented on Figure 7

Peak #	1 h			5h		
	Pos. [°2Th.]	Rel. Int. [%]	FWHM [°2Th.]	Pos. [°2Th.]	Rel. Int. [%]	FWHM [°2Th.]
1	27,0969	86,01	0,551	27,0321	94,12	1,2595
2	36,1845	84,7	0,7085	36,0696	92,65	0,3936
3	54,1691	100	0,9446	54,4088	100	1,1021

Conclusions

Coarse aggregated crystalline particles of rutile with sizes ranging from 2 to 6μm were successfully synthesized by applying forced hydrolysis to 0.94M TiCl₄ solution. Particle size and SEM imaging analyses gave evidence of the formation and growth of these particles to comprise three different stages: induction and nucleation of crystallites of an average size of 7nm; these unit crystallites undergo fast primary aggregation to form spherical particles of a diameter ranging from 1 to 2μm; finally, bigger non spherical aggregates with size up to 10μm are formed through a secondary aggregation stage.

This study made clear that the initial concentration of the solutions as well as the temperature at which the isothermal treatment proceeds has a considerable impact on the stability of the solutions as well as on the kinetics of TiCl₄ hydrolysis. It was found that the induction period increases with the initial concentration of the solutions and decreases with the elevation of temperature. Modelling of the experimental data according to the JMAK equation in the case of isothermal transformation of homogeneous solutions was in agreement with the SEM imaging analyses of the synthesized particles.

Acknowledgments

CC is the recipient of the Werner Graupe Fellowship for which she is grateful to McGill University. The research is funded through a NSERC discovery grant.

References

[1] I. Ilisz, A. Dombi, K. Mogyorósi and I. Dékány, "Photocatalytic water treatment with different TiO2 nanoparticles and hydrophilic/hydrophobic layer silicate adsorbents," *Colloids and Surfaces A: Physicochemical and Engineering Aspects*, 230 (1-3) (2003), 89-97.

[2] J. M. Coronado, S. Kataoka, I. Tejedor-Tejedor and M. A. Anderson, "Dynamic phenomena during the photocatalytic oxidation of ethanol and acetone over nanocrystalline TiO2: simultaneous FTIR analysis of gas and surface species," *Journal of Catalysis*, 219 (1) (2003), 219-230.

[3] S. -H. Lee, S. Pumprueg, B. Moudgil and W. Sigmund, "Inactivation of bacterial endospores by photocatalytic nanocomposites," *Colloids and Surfaces B: Biointerfaces*, 40 (2) (2005), 93-98.

[4] M. Koelsch, S. Cassaignon, C. Minh, J. -F. Guillemoles and J. -P. Jolivet, "Electrochemical comparative study of titania (anatase, brookite and rutile) nanoparticles synthesized in aqueous medium," *Thin Solid Films*, 451-452 (2004), 86-92.

[5] J. N. Hart, R. Cervini, Y. -B. Cheng, G. P. Simon and L. Spiccia, "Formation of anatase TiO2 by microwave processing," *Solar Energy Materials and Solar Cells*, 84 (1-4) (2004), 135-143.

[6] P. V. A. Padmanabhan, K. P. Sreekumar, T. K. Thiyagarajan, R. U. Satpute, K. Bhanumurthy, P. Sengupta, G. K. Dey and K.G.K. Warrier, "Nano-crystalline titanium dioxide formed by reactive plasma synthesis," *Vacuum*, 80 (11-12) (2006), 1252-1255.

[7] K. Baba and R. Hatada, "Synthesis and properties of TiO_2 thin films by plasma source ion implantation," *Surface and Coatings Technology*, 136 (1-3) (2001), 241-243.

[8] D. Zhe, X. Hu, P. L. Yue, L. Q. Gao and P. F. Greenfield, "Synthesis of anatase TiO2 supported on porous solids by chemical vapor deposition," *Catalysis Today*, 68 (1-3) (2001), 173-182.

[9] K. Bok-Hee, L. Jo-Young , C. Yong-Ho, M. Higuchi and N. Mizutani, "Preparation of TiO_2 thin film by liquid sprayed mist CVD method," *Materials Science and Engineering B*, 107 (3) (2004), 289-294.

[10] D. Verhulst, B. J. Sabacky, T. M. Spitler, J. Prochazka, "A new process for the production of nano-sized TiO2 and other ceramic oxides by spray hydrolysis," Nanotechnology & PM2: Scientific Challenges & Commercial Opportunities, Proceedings of the International Conference on Nanotechnology & PM2: Scientific Challenges & Commercial Opportunities, Providence, RI, United States, Sept. 17-18, 2003 (2003), 142-149.

[11] S. Gablenz, D. Voltzke, H. -P. Abicht, J. Neumann-Zdralek, "Preparation of fine TiO2 powders via spray hydrolysis of titanium tetraisopropoxide," *Journal of Materials Science Letters*, 17(7) (1998), 537-539.

[12] C. Wang, Q. Li and R. -D. Wang, "Synthesis and characterization of mesoporous TiO2 with anatase wall," *Materials Letters*, 58 (9) (2004), 1424-1426.

[13] S. J. Bu, Z. G. Jin, X. X. Liu, L. R. Yang and Z. J. Cheng, "Synthesis of TiO_2 porous thin films by polyethylene glycol templating and chemistry of the process," *Journal of the European Ceramic Society*, 25 (5) (2005), 673-679.

[14] S. –J. Kim, S. –D. Park, Y. H. Jeong, S. Park, "Homogeneous precipitation of TiO2 ultrafine powders from aqueous $TiOCl_2$ solution," *Journal of the American Ceramic Society*, 82 (4) (1999), 927-932.

[15] S. –D. Park, Y. H. Cho, W. W. Kim, S. –J. Kim, "Understanding of homogeneous spontaneous precipitation for monodispersed TiO_2 ultrafine powders with rutile phase around room temperature," *Journal of Solid State Chemistry*, 146 (1) (1999), 230-238.

[16] J. H. Lee and Y. S. Yang, "Effect of hydrolysis conditions on morphology and phase content in the crystalline TiO2 nanoparticles synthesized from aqueous $TiCl_4$ solution by precipitation," *Materials Chemistry and Physics*, 93 (1) (2005), 237-242.

[17] H. Xie, Q. Zhang, T. Xi, J. Wang, Y. Liu, "Thermal analysis on nanosized TiO2 prepared by hydrolysis," *Thermochimica Acta*, 381 (1) (2002), 45-48.

[18] Q. Zhang, L. Gao and J. Guo, "Effect of hydrolysis conditions on morphology and crystallization of nanosized TiO_2 powder," *Journal of the European Ceramic Society*, 20 (12) (2000), 2153-2158.

[19] L. Gao, Q. Zhang, "The promoting effect of sulphate ions on the nucleation of TiO_2 (Anatase) nanocrystals," *Materials Transactions*, 42 (8) (2001), 1676-1680.

[20] Y. Hu, H. -L. Tsai and C. -L. Huang, "Phase transformation of precipitated TiO_2 nanoparticles," *Materials Science and Engineering A*, 344 (1-2) (15 March 2003), 209-214.

[21] T. Kapias, R. F. Griffiths, "Accidental releases of titanium tetrachloride (TiCl4) in the context of major hazards-spill behavior using REACTPOOL," *Journal of Hazardous Materials*, 119 (1-3) (2005), 41-52.

[22] E. Rolia, J. E. Dutrizac, "The determination of free acid in zinc processing solutions," *Canadian Metallurgical Quarterly*, 23 (2) (1984), 159-167.

[23] E. Matijevic and R.S. Sapiesko, "Forced Hydrolysis in Homogeneous Solutions" in Fine Particles, Synthesis, Characterization and Mechanisms of Growth, T. Sugimoto Ed., vol. 92 (2000), 2-31.

[24] F. A. Cotton and G. Wilkinson, Advanced Inorganic Chemistry (Fifth edition), Wiley Interscience Ed. (1988), 654-659.

[25] I. Cservenyak, G. H. Kelsall, W. Wang, "Reduction of Ti(IV) species in aqueous sulfuric and hydrochloric acid I. Titanium speciation," *Electrochimica Acta*, 41 (4) (1996), 563-72.

[26] M.C. Weinberg, D.P. Birnie III, V.A. Shneidman, "Crystallization kinetics and the JMAK equation," *Journal of Non-Crystalline Solids*, 219 (1997), 89-99.

[27] M. Avrami, "Kinetics of Phase Change. I General Theory," *J. Chem. Phys.*, 7 (1939), 1103-1112.

[28] M. Avrami, "Kinetics of Phase Change. II Transformation-Time Relations for Random Distribution of Nuclei," *J. Chem. Phys.*, 8 (1940), 212-221.

[29] R. Sarraf-Mamoory, G.P. Demopoulos, R.A.L. Drew, "Preparation of fine copper powders from organic media by reaction with hydrogen under pressure; Part II The kinetics of particle nucleation, growth and dispersion," *Metallurgical and Materials transactions B*, 27B (1996), 585-594.

[30] R. M. Cornell and U. Schwertmann, "The Iron Oxides, Structure, Properties, Reactions, Occurence and Uses, VCH Ed., New York (1996), Peak Width – Crystal Size and Strain, 38-39.

PRODUCING TITANIUM BY TITANIA ELECTROLYSIS IN A 5KA CELL

Huimin Lu[1], Huanqing Han[1]

[1]Beijing Univ. of Aeronautics & Astronautics, School of Material Sci. & Eng.;
37 Xueyuan Road, Haidian District, Beijing 100083, China

Keywords: TiO_2 electrolysis, current efficiency, energy consumption

Abstract

In this paper, the theoretical energy consumption of TiO_2 electrolysis is calculated and the real energy consumption of TiO_2 electrolysis is measured in a 5kA electrolysis cell. In the experiments of TiO_2 electrolysis, some important measures like the modification of cell design and ultrasonic electrolytic technology are adopted for improving the current efficiency. This process produces titanium with 0.38%O, 0.032%C, 0.0090H, 0.061N and 0.12%Fe, the current efficiency range is 70%~82%, the energy consumption is 7.984kWh/kg Ti and carbon dioxide and carbon monoxide are discharged on carbon anode. This new technique has low energy consumption, high current efficiency and less pollution for the environment. The reduction mechanism of TiO_2 to Ti was also studied.

Introduction

Titanium and its alloys are attractive engineering materials because of their lightness, strength, and corrosion resistance. However, unlike common metals, such as steel, aluminum alloys, zinc, lead, copper, etc., titanium and its alloys only find applications in restricted areas such as airplanes and golf clubs. This is not because of the rarity of titanium, which is the fourth most structural metal in the earth's crust after aluminum, iron and magnesium [1], but its high processing cost. So far, The Kroll process still is the only titanium sponge production method in industry. The Kroll process consists of a three-step operation of TiO_2 to $TiCl_4$, $TiCl_4$ to sponge Ti by Mg liquid and the electrochemical recycling of $MgCl_2$ into metallic Mg. It takes 2 to 5 day in this reduction route via $TiCl_4$, and the batch operation makes it difficult to save thermal energy [2, 3]. The process has been widely used over the last 50 years for the commercial production of titanium metal. Some new processes for production of titanium sponge uninterruptedly came into being, however, so far no a new process has been applied in industry [4]. The FFC Cambridge process was invented by Professor Derek J. Fray and his team in 1997 mainly for reduction of TiO_2 to Ti sponge at first [5]. It is a kind of new process of short flow and low consume of energy to produce sponge Ti instead of the traditional Kroll process. The simpler, more rapid, and compact process in a single step directly from TiO_2 has been desired to achieve higher productivity and energy saving [6~8]. We conducted some experiments on the FFC Cambridge process in laboratory. These results showed that the FFC Cambridge process had lower current efficiency about 30%~50%. It was likely that some parasitic reactions such as carbon precipitation, the lower O^{2-} diffuse transmitting rate in the molten $CaCl_2$, the TiO_2 plate cathode used and the back-reaction lowered the current efficiency. The process need improve so as to be applied in industry.

In this paper, we report the theoretical energy consumption of TiO_2 electrolysis and the real energy consumption of TiO_2 electrolysis, some important measures such as the modification of cell design, ultrasonic electrolytic technology adopted for improving the current efficiency in a 5kA electrolysis cell.

Energy consumption

Theoretical energy consumption in titanium electrolysis is composed of two parts: (a) the energy required for heating titanium dioxide and anode graphite; (b) the energy required for decomposing titanium dioxide. The reaction is as follow:

$$TiO_2(s,298K) + \frac{2}{3}(1+\frac{1}{x})C(s,298K) = Ti(s,TK) + \frac{2}{3}(2-\frac{1}{x})CO_2(g,TK) + \frac{2}{3}(\frac{2}{x}-1)CO(g,TK) \quad (1)$$

where x – current efficiency.
The energy for the above reaction is

$$H_{total} = (0.507 + 3.720 \times 10^{-3}T + 9.783 \times 10^{-8}T^2 + 69.893\frac{1}{T})$$
$$+ x^{-1}(0.333 - 2.744 \times 10^{-4}T - 1.488 \times 10^{-8}T^2 + 91.564\frac{1}{T}](kWh/kgTi) \quad (2)$$

when x = 100%,

$$H_{total} = [0.84 + 3.446 \times 10^{-3}T + 8.295 \times 10^{-8}T^2 + 1.615 \times 10^2\frac{1}{T}](kWh/kgTi) \quad (3)$$

when x = 70%,

$$H_{total} = [0.983 + 3.328 \times 10^{-3}T + 7.657 \times 10^{-8}T^2 + 2.007 \times 10^2\frac{1}{T}](kWh/kgTi) \quad (4)$$

The dependence of energy consumption on temperature is shown in Table 1.

Table 1   The dependence of energy consumption (kWh/kg Ti) on temperature (T)

T/K	1073	1098	1123	1148	1173	1198	1223	1248	1273
x=100%	4.784	4.871	4.958	5.046	5.134	5.222	5.311	5.399	5.488
x=70%	4.829	4.912	4.996	5.079	5.165	5.247	5.332	5.418	5.501

From the above relationships, it can be seen clearly that the theoretical energy consumption decrease with increasing current efficiency and with decreasing temperature. At 1173K the theoretical energy consumption would be 5.134kWh/kg Ti at current efficiency 100%. It decreases as temperature decreases. The energy gradient is

$$\frac{d\Delta H_T^0}{dT} = -3.523 \times 10^{-3}kWh/kgTi \quad \text{(at 100\% CE)} \quad (5)$$

Experimental

Electrolysis of TiO$_2$ in 5kA electrolysis cell

Electrolysis was carried out in a 5kA electrolysis cell, which is shown in Fig. 1. A graphite crucible was used, which served both as the container in which CaCl$_2$ (>99pct) and LiCl (>99pct) melt electrolyte was and as the anode, which density was 1.7g/cm^3. A cylindrical basket-type steel container (ϕ16cm×H 35cm) was used as the cathode in which the about 14kg spherical TiO$_2$ (>99pct, rutile, primary particle size <1μm, spherical particles diameter 1cm) was filled. The diameter 3mm pores spread all over cylindrical surface. The anode and cathode was isolated by a pure Ti net with -120 mesh pores. The anode-cathode distance was 40mm. The duration of electrolysis was 4h. The current efficiency (CE) was determined from the amount of oxygen losing by weighing after the experiments.

Before electrolysis of TiO_2, the pre-electrolysis with a stainless steel cathode rod at 1.0 to 1.3 V was conducted at 1173 K to eliminate metallic impurities and water. After pre-electrolysis, the basket-type cathode was immersed in the salt (about 30cm in depth) and held at 30 mm above the bottom at 1173 ± 5 K. A direct current was supplied under a constant voltage 2.8V.

In electrolysis, the ultrasonic electrolytic technology was applied. The work frequency range and electro-power of the type SCQ250 ultrasonoscope were 20 kHz ~60 kHz and 2000W respectively. After electrolysis, the cathode was taken out of the melt and cooled at the upper part of the furnace. The solidified salt in the basket-type cathode was leached in the flow of drinking water. The black Ti powder was subsequently rimmed with dilute HCl, distilled water in that order, and then dried in vacuum for analysis. The morphology was observed using scanning electron microscopy (SEM) equipped with an energy-dispersive X-ray (EDX) analyzer.

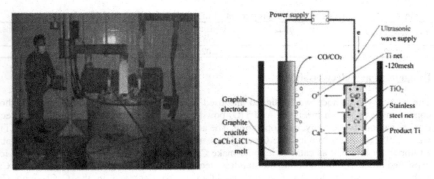

Fig.1 5kA electrolysis cell and schematic illustration of the 5kA electrolysis cell
for TiO_2 reduction

Electrochemical study on reaction mechanism

The 400-800 size abrasive paper was used to sand surface of Ti (commercially pure, 60mm × 60mm × 1mm), acetone to get rid of the oil, ultrasound to clean the Ti in ethyl alcohol. The cleaned Ti was hung on a crucible kept temperature at 700°C in air for 10 day. TiO_2 film appearing red-gray color was formed on the surface of Ti.

Dried and analytically pure $CaCl_2$ was put into graphite crucible in tube type resistance furnace heated to required temperature in Ar atmosphere under control of temperature controller. Ti and TiO_2 as work electrodes, graphite as negative electrode, graphite crucible as the reference electrode, molybdenum wire as leading wire, all of these above made up of electro-chemical measure system. The electro-chemical measurements were conducted by Solartron1480 electrode potentials gauge made in UK.

Results and discussion

Adding LiCl to the molten $CaCl_2$ electrolyte

99

The physicochemical properties of the molten electrolyte are important both from theoretical and technical point of view, because they are related to the structure of molten salts and the current efficiency in the electrolytic process. In the experiments, add 5 mass% LiCl to CaCl$_2$ melt. LiCl additive can lower the liquidus temperature (the melting point) of CaCl$_2$. Decrease in liquidus temperature is generally considered advantageous, since the operating temperature of electrolysis may then become possible to be lowered, and this improves the cell performance of the process. The electrical conductivity of the electrolyte is a physicochemical property of major concern, since it is directly related to the energy consumption of the process. The ohmic voltage drop in the electrolyte contributes to about one third of the total cell voltage loss. The viscosity of the electrolyte influences the hydrodynamic processes in the electrolysis cell, e.g. O^{2-} diffuse rate and the release of O^{2-} from the cathode surface. A high viscosity will lower the O^{2-} diffuse rate and the release of O^{2-} from the cathode surface. Therefore, we selected LiCl as additive because LiCl has higher decomposition voltage than CaCl$_2$ and can improve the physicochemical properties of the molten CaCl$_2$. In experiments, we added 5 mass% LiCl to the molten CaCl$_2$. The physicochemical properties of the molten electrolyte were shown in Table 2.

Table 2   The physicochemical properties of the molten CaCl$_2$ + 5%LiCl electrolyte

Liquidus temp./ °C	Electrical cond./ $\Omega^{-1} \cdot m^{-1}$	Density/ g·cm^{-3}	Viscosity/ Pa·s	Interfacial tens./ J·m^{-2}
602	12.1	2.02	1.822	309

Designing a new electrolysis cell

The basket-type cathode was designed and used in the experiment for increasing the cathode conducting area shown in Fig. 1. And TiO$_2$ powder put in the basket-type cathode and directly used as cathode was simply, easy operation. The anode and cathode was isolated by a pure Ti net with -120 mesh pores to eliminate C parasitic reactions with Ti. The stainless steel basket plaid an important role that as cathode it could make CaO discompose in Ca^{2+} and O^{2-}, the Ca^{2+} ions deposited on the cathode and into the stainless steel basket for reacting with TiO$_2$ producing CaO and Ti. Because of the molten CaCl$_2$ could dissolved CaO and in the CaCl$_2$ melt CaO concentration decreased uninterruptedly by the CaO electrolysis, the new producing CaO would enter the CaCl$_2$ melt owing to the concentration gradient force. Therefore, calcium can reduce TiO$_2$ to Ti in the process. In the meantime, the TiO$_2$ cathode in the stainless steel basket can deoxygenate electrochemically to form O^{2-} and Ti^{4+}, the O^{2-} diffuses and moves towards anode, the Ti^{4+} gets 4 electrons from cathode to form metal Ti on the spot. We think there are two reactions in the process, i.e. calcium reduction of TiO$_2$ and TiO$_2$ deoxygenating electrochemically. The new electrolysis cell design was beneficial to increase current efficiency.

Applying ultrasonic electrolytic technology

The FFC Cambridge process is firstly ionization of the oxygen in the TiO$_2$ cathode, the ionized O^{2-} diffuses and moves towards anode, then the ionized O^{2-} and C anode chemical reaction occurs. In the process, the ionized O^{2-} diffusion is an important step. If the O^{2-} diffusion rate is lower, the TiO$_2$ electrolysis to Ti will take a long time, which results in the low current efficiency.

The ultrasonic work frequency range is 20 kHz ~60 kHz, its intensification lies in the mechanical effect[9]. When the ultrasonic wave disseminates in the melt, the changing pressure producing in the process will stir the melt, thereby markedly lower the diffusion resistance. In the experiments, the optimum ultrasonic work frequency range is 40~45 kHz.

Comprehensive experiments

In the molten $CaCl_2$ molten electrolyte with 5 mass% LiCl, the basket-type cathode using stainless steel as the cathode material was designed and used, the electrolysis operating voltage was 2.8V, the anode is the graphite crucible, the ultrasonic work frequency applied was 42kHz, the electrolysis time was 4h, the current efficiency was 70~82%. Fig.2 is the cathode products image, the SEM images of Ti powder and XRD analysis from TiO_2 electrolysis. The chemical composition of cathode products was as follows: 0.38%O, 0.032%C, 0.0090H, 0.061N and 0.12%Fe.

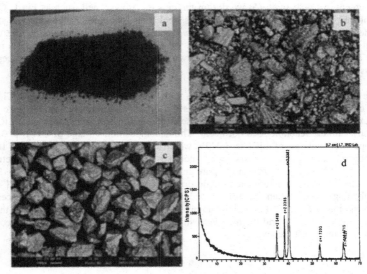

Fig.2 The images of the cathode products Ti powder (a), the SEM (b)
(c) and XRD analysis(d)

Electrochemical study on reaction mechanism

We think in our TiO_2 electrolysis process, there are two reactions occurred in the same time. The two process reactions are FFC Cambridge process and OS process.

FFC Cambridge Process :

$$\text{Cathode: } 3TiO_2 + 12e^- \leftrightarrow 3Ti + 6O^{2-} \tag{6}$$
$$\text{Anode: } 4C + 6O^{2-} \leftrightarrow 2CO_2 + 2CO + 12e^- \tag{7}$$
$$\text{Overall reaction: } 3TiO_2 + 4C \leftrightarrow 3Ti + 2CO_2 + 2CO \tag{8}$$

OS Process:

$$\text{Cathode: } 6Ca^{2+} + 12e^- \leftrightarrow 6Ca \tag{9}$$
$$3TiO_2 + 6Ca \leftrightarrow 3Ti + 6CaO \tag{10}$$
$$\text{Anode: } 4C + 6O^{2-} \leftrightarrow 2CO_2 + 2CO + 12e^- \tag{11}$$
$$\text{Overall reaction: } 3TiO_2 + 4C \leftrightarrow 3Ti + 2CO_2 + 2CO$$

101

In OS experiments, first adding CaO to the CaCl₂ melt, let the CaO electrolysis for producing Ca, make Ca reduces the TiO₂.

In our experiment, (1) through baking TiO₂ above 1200°C, the TiO₂ can change into TiO$_x$ form, the TiO$_x$ has better conducing electricity; (2) when using TiO$_x$ as cathode, the cathode can have better conducing electricity; (3) when applying a volt on the cathode and anode, first an electrochemical reaction occurs at the cathode, O^{2-} ionizes and moves to anode, in the process meets Ca^{2+} and forms CaO, it resolves in CaCl₂ melt; (4) when there are much CaO resolved in CaCl₂ melt, the electrolysis effect of TiO₂ is not good. Therefore, it is important to reduce the CaO content in CaCl₂ melt. (5) we design the new electrolysis cell, such as Fig.1, CaO can be electrolyzed between anode and the stainless steel basket cathode; in the stainless steel basket cathode, there is a lead connecting the power cathode, it takes a important role conducting current to TiO$_x$, make the TiO$_x$ ionized, and the Ca formed from CaO electrolysis reduces the TiO$_x$. Therefore, in our process there are two process i.e. FFC Cambridge process and OS process.

On the LiCl role when adding LiCl to CaCl₂ melt, we think it only makes the CaCl₂ melt have better conducting electricity. Fig. 3(a) shows the cyclic voltammogram in the pure anhydrous pure LiCl melt. From this Figure it can be seen that in the pure anhydrous pure LiCl melt the TiO₂ any cathode reaction does not occur. The decomposition voltage of CaCl₂ is 3.213V; the decomposition voltage of LiCl is 3.242V when the melt temperature is 1173K. If the TiO₂ electrochemical reaction on cathode has nothing to do with Ca^{2+} ions, the electrochemical reaction should occur similarly in the pure anhydrous pure CaCl₂ melt as Fig.3 (b). In fact, the TiO₂ electrochemical reaction on cathode does not occur, which shows that LiCl only plays strengthening CaCl₂ molten electrolyte conductivity role, at the same time, it is necessary TiO₂ deoxygenating only in CaCl₂ melt. Adding 5% LiCl to CaCl₂ melt can lower the liquidus temperature about 175K and the melt viscosity, it is beneficial to lowering energy consumption and raising current efficiency.

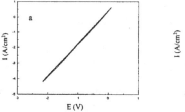 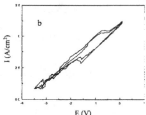

Fig. 3 Cyclic voltammogram in pure anhydrous pure LiCl melt (a) or pure anhydrous pure CaCl₂ melt (b) at 1173K, scanning rate: 10 mV/s, working electrode area: 0.475cm², TiO₂ electrode being preparing by baking Ti plate at 1273K

Fig. 4 is Chronoamperometric response curves of TiO₂ electrode at 1173K. Take current values at different potential stage at the same time to make I(t)-E curve, as shown in Figure 5. When the potential stage is negative enough to reach diffusion control potential, a flat appears on I (t)-E curve by which the extreme current can be calculated. With regard to a simple electron transmission reversible reaction, the following equation is tenable when product is dissolving[10].

102

$$E = E_{\frac{1}{2}} + \frac{RT}{nF}\ln[\frac{I_d - I_t}{I_t}]$$ (12)

Where, E, $E_{1/2}$, I_d, I_t, n are the reduction electron potential, semiwave potential, extreme current, current intensity of t time and electron number of electrode reduction transmission. The E vs. ln $[(I_d - I_t)/I_t$ figure is made. We can get tow lines. According to the rate of slope k = RT/nF, n is calculated about 2. It means that the gain and loss electron number during the redox reaction on the TiO_2 electrode when potential is –0.46V and –1.78V is 2. We can determine that the reaction is $TiO_2 + 2e = TiO + O^{2-}$ on the TiO_2 electrode when potential is –0.46V and When the potential reaches –1.78V, then the reaction $TiO + 2e = Ti + O^{2-}$ occurs.

Washing the electrode by water at room temperature after the experiments ended. We caught observe there were some metal distributed on the surface. The results by SEM and element analytical showed that part of TiO_2 electrode had changed into Ti.

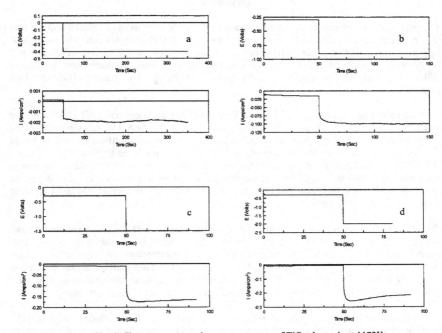

Fig. 4 Chronoamperometric response curves of TiO_2 electrode at 1173K:
(a) -0.4V;(b)-0.7V; (c)-1.5V; (d)-2.0V

103

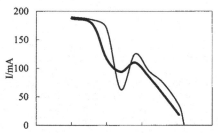

Fig. 5 Relationship between the current and the voltage of the
TiO$_2$ electrode at 1173K

Conclusions

In this paper, the theoretical energy consumption of TiO$_2$ electrolysis is calculated and the real energy consumption of TiO$_2$ electrolysis is measured in a 5kA electrolysis cell. The theoretical energy consumption of the TiO$_2$ electrolysis decreases with increasing current efficiency and with decreasing temperature. At 1173K the theoretical energy consumptions would be 5.134kWh/kg Ti at current efficiency 100% and 5.165kWh/kg Ti at current efficiency 70% respectively.

The direct reduction experiments of TiO$_2$ to Ti in molten CaCl$_2$ and 5% LiCl electrolyte are conducted for improving the process current efficiency in the 5kA electrolysis cell. In the experiments of TiO$_2$ electrolysis, some important measures like the modification of cell design and ultrasonic electrolytic technology are adopted. This process produces titanium with 0.38%O, 0.032%C, 0.0090H, 0.061N and 0.12%Fe, the current efficiency range is 70%~82%, the energy consumption is 7.984kWh/kg Ti and carbon dioxide and carbon monoxide are discharged on carbon anode. This new technique has low energy consumption, high current efficiency and less pollution for the environment.

In the TiO$_2$ electrolysis process, there are two reactions which are FFC Cambridge process and OS process occurred in the same time. Therefore, the process is a combining process with deoxygenating and calcium reduction reactions. The TiO$_2$ electrochemical reaction on cathode does not occur in pure LiCl melt, LiCl only plays strengthening CaCl$_2$ molten electrolyte conductivity role. Adding LiCl to CaCl$_2$ melt can lower the liquidus temperature and the melt viscosity, it is beneficial to lowering energy consumption and raising current efficiency.

The deoxygenating process is firstly ionization of the oxygen in the TiO$_2$ cathode, the ionized O^{2-} diffuses and moves towards anode, then the ionized O^{2-} and C anode chemical reaction occurs. In the process, the ionized O^{2-} diffusion is an important step. The O^{2-} diffusion rate is lower; the TiO$_2$ electrolysis to Ti takes a long time, which results in the low current efficiency. Applying ultrasonic electrolytic technology can markedly lower the O^{2-} diffusion resistance, raising current efficiency.

Calculation of timing-current curve of TiO$_2$ electrode in CaCl$_2$ flux shows that reduction of TiO$_2$ is conducted in two steps, firstly TiO$_2$ is reduced to TiO, and then TiO is reduced to Ti. The reduction potentials in both steps are decreased with the increase of the temperature.

References

1 F. Habashi, handbook of extractive metallurgy, 1997, Weinheim, Germany, Wily-VCH.

2 W. Kroll. Trans. Electrochem. Soc., 1940, 78, 35 – 47.

3 Honggui Li. Rare Metal Metallurgy (in Chinese), Beijing, Metallurgical Industry Press, 2001, 104.

4 Kang Sun. Physical-chemistry of Titanium Exaction (in Chinese), Beijing, Metallurgical Industry Press, 2001, 234.

5 George Z. Chen, Dere J. Fray and Tom W.Farthing. Nature, Vol407, 21(2000), 361.

6 Gaorge Zheng Chen, Derek J. Fray and Tom W. Farthing. Direct Electrochemical Reduction of Titanium Dioxide to Titanium in Molten Calcium Chloride. Nature, 2000, 9(407), 361.

7 Harvey M. Flower. A Moving Oxygen Story. Nature, 2000, 9(407), 305.

8 Gaorge Zheng Chen, Derek J. Fray and Tom W. Farthing. Metall. Mater. Trans. B, 2001, vol.32 B (6), 1041 – 1052.

9 Shiguang Chen. Application of Ultrasound in Hydrometallurgical Processes. Shanghai Nonferrous Metals (in Chinese), 2003, 24(3), 142 – 146.

10 Zukang Zhang, Erkang Wang. Electrochemical Principle and Method (in Chinese). Beijing, Science Press. 2000, 55.

Innovations in
Titanium
Technology

Novel Materials and Processes II

PROCESSING ROUTES TO PRODUCE TITANIUM FROM TiO$_2$

Dr. J. C. Withers [1], J. Laughlin, and Dr. R. O. Loutfy
[1]MER Corporation, 7960 S. Kolb Road, Tucson, AZ 85706, U.S.A.

Keywords: electrowinning titanium, titanium from its oxide, carbothermic reduction of TiO$_2$

Abstract

The only commercial process to produce titanium metal is the metallothermic reduction of TiCl$_4$ by sodium (Hunter) and magnesium (Kroll). If titanium could be produced from the ore/TiO$_2$ a saving for producing TiCl$_4$ and its special storage and handling would at least be achieved as well as the electrolytic requirement of producing the Na or Mg reductant. An approach is to carbothermically reduce TiO$_2$ to a lower oxide which provides several possible routes for a second electrolytic reduction step to produce titanium metal. Several potential electrolytic reduction routes for producing titanium from a carbothermic produced suboxide will be discussed with experimental results provided for each process route.

Background

Titanium is produced from the metallothermic reduction of titanium tetrachloride (TiCl$_4$, tickle). Primarily magnesium is used known as the Kroll process. One operation utilizes sodium as the reductant known as the Hunter process. The TiCl$_4$ is produced from chlorinating ore directly and purifying the chlorides through vacuum distillation or from a hydrometallurgical purified ore to TiO$_2$ which is then chlorinated. The production of purified TiCl$_4$ and electrolytical production of magnesium or sodium, and the metallothermic reduction results in high cost primary titanium. Although the Kroll and Hunter processes have been in practice for over 50 years, no other process has emerged that can commercially produce titanium at lower cost. Most alternative processing has utilized electrolysis employing the readily available TiCl$_4$ as feed. In spite of extensive wide spread electrolysis investigation no processing has demonstrated sufficient success to commercially challenge the more expensive Kroll process.

Reducing the cost of titanium is enabling for titanium's application in virtually all applications which prompted DARPA to create an Initiative in Titanium whose purpose was to reduce the cost of primary titanium to substantially less than Kroll produced primary sponge. An approach to reduce cost is to eliminate the use of TiCl$_4$/tickle as a feed and utilize TiO$_2$. Aluminum is economically produced from its oxide and comparatively the energy to extract titanium from its oxide is slightly less than aluminum from its oxide which suggests theoretically utilizing TiO$_2$ as a feed has potential to economically produce titanium. Unfortunately, no analogous system of dissolving alumina (Al$_2$O$_3$) in cryolite and electrolyzing to produce pure aluminum has been found for dissolving TiO$_2$ and producing titanium free of oxygen. The Fray/FFC–Cambridge process utilizes TiO$_2$ as a cathode [1-3] and reportably can extract oxygen and produce titanium. This was one of the Phase I processes sponsored by DARPA to produce reduced cost primary titanium. Through several years research by several investigators, no pilot or commercial production process has emerged utilizing TiO$_2$ as a cathode to produce low oxygen content primary titanium. Uno [4-9] and others [10-11] have reported electrolytically depositing calcium that reduces TiO$_2$ and the product CaO is utilized to regenerate calcium in electrolysis of CaO with a graphite anode. The produced titanium has not been reported to meet the specifications equivalent to Kroll sponge and there are no reports of the process being scaled-up.

Under a DARPA and Army SBIR program MER has investigated utilizing TiO_2 or an ore containing TiO_2 to produce titanium. The concept was to utilize TiO_2 and carbon with an initial heat treatment followed by electrolysis to complete the reduction and produce titanium.

Processing Routes to Produce Titanium from TiO_2

These are several possible processing routes to produce titanium from a TiO_2 source. Some of the possible reaction routes are illustrated in Figure 1. As can be seen in each case a carbon source is mixed with TiO_2 and heated to produce an intermediate product which severs as the feed for electrolysis to electrowin titanium.

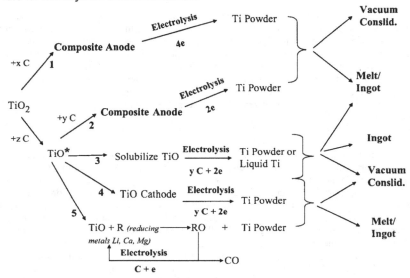

* TiO represents a suboxide which may not be pure TiO and may contain residual carbon in solid solution.

Figure 1: Different Processing Routes to Produce Ti from TiO_2 Feed

In the top most reaction route 1, the heat treatment temperature of TiO_2 and carbon is maintained below any carbothermic reduction at approximately 1100 °C and the composite anode is used for electrolysis to electrowin titanium. Because of the multivalences of titanium disproportionation can occur between Ti^{+4}, Ti^{+3}, Ti^{+2}, and $Ti°$ to result in poor columbic efficiency even though it is possible to produce titanium particulate from such a composite anode in which the titanium is in a +4 valence state.

In reaction route 2 the TiO_2 and carbon are heat treated to a sufficient temperature to produce a titanium suboxide which reduces the titanium to a +2 valence. The oxygen content may be equal or less than stoichiometric TiO. There is sufficient carbon used, designated as z, to provide CO and/or CO_2 during subsequent electrolysis, or just enough to produce TiO as an intermediate. The TiO is then mixed with stoichiometric carbon that an electrolysis will produce CO and/or CO_2 and titanium particulate. Producing a titanium suboxide and composite anode that is electrolyzed to titanium is presented in a separate paper at his symposium.

110

In reaction route 3, the suboxide produced is used as a soluble feed to a fused salt. Graphite is used as an anode that on electrolysis titanium particulate is produced with CO at the anode. This reaction route de-couples stoichiometry between the suboxide and the graphite anode as the system is self adjusting based on the solubility of the suboxide in the fused salt electrolyte and the current utilized for electrolysis. If the electrolysis operation is performed below the melting point of titanium, a particulate morphology is produced. Electrolysis has been performed above the melting point of titanium which produces the titanium in a molten form.

In reaction route 4 the carbothermic produced suboxide can be used cathodically from which oxygen can be extracted using a graphite anode that discharges CO or inert dimensionally stable anode that discharges oxygen. The suboxide is very electrically conductive and takes less electrons to remove the remaining low oxygen content as well as is less susceptible to parasitic reactions such as titanates being formed that reduces columbic efficiency.

In reaction route 5, the carbothermic produced suboxide is also used cathodically, but in this case a reducing metal whose oxide is more thermodynamically stable than TiO_2 is deposited at the cathode to reduce the titanium suboxide and form the reducing metal oxide. The reducing metal oxide is electrolyzed using a graphite anode that discharges CO or inert dimensionally stable anode that discharges oxygen.

Results of Investigations

Others [12-13] have reported the investigation of producing titanium from a TiO_2-C anode in which the heat treatment temperature of the TiO_2-C is below where carbothermic reduction begins which is approximately 1000-1100 °C. In this investigation stoichiometric TiO_2 and C anodes were molded and heat treated to 1100 °C. The anodes were electrolyzed in salt compositions of $AlCl_3$-NaCl eutectic at 180 °C, the $AlCl_3$-NaCl eutectic with 20 mole % NaF at 400 °C and the eutectic of NaCl-$CaCl_2$. Short term electrolysis at 0.25 and 1 amp/cm^2 produced a particulate on the cathode that was identified as titanium using EDS in a SEM.

Reaction 2 Investigations an Reported Separately at this Symposium

In reaction route 3 the carbothermic reduced suboxide was identified primarly as TiO as shown in the XRD Figure 2. This intermediate suboxide was utilized as a feed in calcium chloride ($CaCl_2$), $CaCl_2$-20 mole % CaF_2 and $CaCl_2$-NaCl. The salt was operated at 850-900 °C using a graphite anode and molybedenum cathode. Short duration electrolysis at current densities of 0.25 and 1 amp/am^2 was performed. The cathode deposits were identified as titanium by EDS in a SEM. The morphology of the deposits were somewhat flakey as shown in Figure 3.

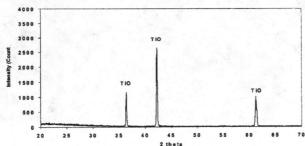

Figure 2: XRD of Carbothermically Reduced TiO2 and C to Produce the Suboxide as Primarily TiO.

(a)

(b)

Figure 3: (a) Deposit with TiO as Solute in $CaCl_2$; (b) Deposit with
TiO as Solute in $CaCl_2$-CaF_2

In the case of reaction route 4, the titanium suboxide was ground to minus 80 plus 200 mesh (177 μ - 74 μ) placed in a tungsten mesh basket and electrolyzed in salts of CaCl$_2$, CaCl$_2$-NaCl, and LiCl. A graphite anode was used and electrolysis was maintained at a potential less than that to deposit the cation of the salt. Initially a high current was registered which decayed in one to a few hours. When the current approached near zero which was typically 4-24 hours, the tungsten basket was retrieved, washed, and the remaining product subjected to EDS analysis and oxygen analysis. A before and after electrolysis photo is shown in Figure 4. It was infrequent the oxygen level was reduced to below several thousand ppm, there was always residual carbon in the titanium and the columbic efficiency was relatively low at under 50%.

In the case of reaction route 5, the same set-up was utilized as in reaction route 4 except the potential was maintained at a sufficient level to deposit Ca or Li. A small amount of approximately 5 wt % CaO was added to the CaCl$_2$ salt and approximately 2 wt % Li$_2$O to the LiCl salt. Others [14-15] have utilized this approach to produce uranium from its oxide. In these investigations it observed there was residual calcium in the remaining titanium which was not readily removed. Also, the oxygen level remained at several thousand ppm. Also, there was always residual carbon in the remaining titanium. The LiCl-Li$_2$O system performed superior to the CaCl$_2$-CaO electrolyte system with lower, but yet over 1000 ppm oxygen and the residual carbon remained in the titanium. The columbic efficiencies were also improved somewhat over the CaCl$_2$-CaO system. The morphology of the reduced titanium oxide was substantially the same as the other cathodically reduced material. The morphology from this investigation is similar to that shown in Figure 4. While extensive investigations were not performed, it appears in this cursory work the lithium system in reaction route 5 has greater potential than the other cathode approaches when using titanium suboxide as the cathode.

Figure 4: Titanium Suboxide Particles Before Cathodic Treatment

113

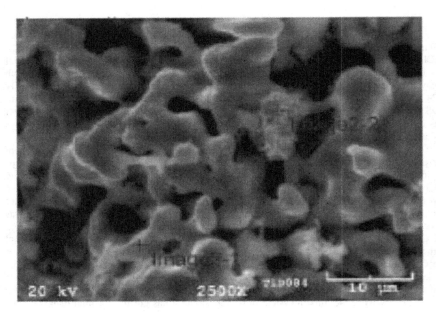

Morphology of Titanium after Cathodic Reduction of a Titanium Suboxide

Conclusions

- There are several possible reaction routes to utilize TiO_2 as a feed with carbon reduction that produces titanium with electrolysis
- TiO_2 can be carbothermically reduced to a suboxide from which titanium can be produced with electrolysis
- Titanium suboxide exhibits sufficient solubility to support electrolysis with a graphite anode in $CaCl_2$ base salts to produce flake morphology titanium deposits
- Titanium suboxide can be used cathodically to electrolytically extract oxygen without titanate formation, but low oxygen titanium was not produced as well as residual carbon contaminated the titanium and poor columbic efficiency was characteristic?
- Calcium and lithium were separately deposited at cathodic titanium suboxide that reduced the titanium suboxide to thousands ppm oxygen, but the produced titanium was contaminated with carbon. The columbic efficiencies were higher than without the cation deposition with the lithium salt system exhibiting the most promise
- Additional work not reported here indicates the carbon contamination can be eliminated.

References

1. G. Z. Chen, D. J. Fray, and T. W. Farthing, Nature, 407 (2000), 361-364.
2. G. Z. Chen, D. J. Fray, *Electro-Deoxidation of Metal Oxides*, Light Metals, TMS 2001.
3. C. Schwandt and D. L. Fray, *Determination of the Kinetics Pathway in the Electrochemical Reduction of Titanium Dioxide in Molten Calcium Chloride*, Electroctimica Acta 51, 2005, 66-76.
4. T. H. Okabe, T. Oishi, and K. Ono, *Preparation and Characterization of Extra-Low-Oxygen Titanium*, Journal of Alloys and Compounds, 184, (1992), 43-56, JALCOM 160.
5. T. H. Okabe, M. Nakamura, T. Oishi, and K. Ono, *Electrochemical Deoxidation of Titanium*, Metallurgical Transactions B, Volume 24B, June 1993, 449.
6. Ryosuke, O. Suzuki, Masayuki Aizawa, Katsutoshi Ono, *Calcium-Deoxidation of Niobium and Titanium in Ca-Saturated $CaCl_2$ Molten Salt*, Journal of Alloys and Compounds, 228, (1999), 173-182.
7. Katsutoshi Ono and Ryosuke O. Suzuki, *A New Concept for Producing Ti Sponge: Calciothermic Reduction*, JOM, February 2002.
8. Ryosuke O. Suzuki and Katsutoshi Ono, *A New Concept of Sponge Titanium Production by Calciothermic Reduction of Titanium Oxide in the Molten $CaCl_2$*, Electrochemical Society Proceedings, Volume 2002-19.
9. Ryosuke O. Suzuki, Koh Teranuma, and Katsutohi Ono, *Calciothermic Reduction of Titanium Oxide and In-Situ Electrolysis in Molten $CaCl_2$*, Metallurgical and Materials Transitions B, Volume 34B, June 2003-287.
10. Ryosuke O. Suzuki and Shuichi Inoue, *Calciothermic Reduction of Titanium Oxide in Molten CaCl2*, Metallurgical and Materials Transaction B, Volume 34B, June 2003-277.
11. Ryosuke O. Suzuki, *Calciothermic Reduction of TiO_2 and In-Situ Electrolysis of CaO in the Molten $CaCl_2$*, Journal of Physics and Chemistry of Solids, 66, (2005), 461-465.
12. Hsin-Yi Haw, Ding-Long Chen, Hann-Woei Tsaur, and Chan-Cheng Yang, *The Electrodeposition of Titanium from the Low Temperature Molten Electrolyte*, Electrochemical Society Proceedings, Volume 99-41.
13. V. Ananth, S. Rajagopan, P. Subramanian, U. Sen, *Single Step Electrolytic Production of Titanium*, Trans. Indian Inst. Met., Vol. 51, No. 5, October 1998, 399-403.
14. Yoshiharu Sakamura, Masaki Kurata, and Tadashi Inoue, *Electrochemical Reduction of UO_2 in Molten $CaCl_2$ or LiCl*, Journal of The Electrochemical Society, 153 (3) D31-D39, (2006).
15. Redey, Laszlo, Williamson, Mark, *Electrochemical Reduction of Metal Oxides in Molten Salts*, Light Metals (Warrendale, PA, United States), 2002, 1075-1082.

"Approved for Public Release, Distribution Unlimited"

115

THE PRODUCTION OF TITANIUM FROM A COMPOSITE ANODE

Dr. J. C. Withers [1], J. Laughlin, and Dr. R. O. Loutfy
[1]MER Corporation, 7960 S. Kolb Road, Tucson, AZ 85706, U.S.A.

Keywords: titanium production, titanium from its oxide, electrowinning titanium

Abstract

The DARPA Initiative in Titanium focused the enabling need for low cost primary titanium which resulted in several DARPA sponsored programs for investigation to produce low cost primary titanium. The composite anode process was one that received DARPA sponsorship. During Phase I, it was demonstrated titanium could be produced in a two-step process. TiO_2 was carbothermically reduced to TiO which was combined in an anode with carbon. The composite anode of TiO-C was electrolyzed in a fused salt to produce titanium particulate at the cathode and CO at the anode. The titanium was separated from the salt by vacuum evaporation and while hot, pressed into billets. The billets were secondarily processing by RMI and the titanium characterized. Processing will be discussed and the properties of titanium presented. This composite anode process is now in Phase II which is scaling-up to 500 lbs/day production of titanium.

Background

It is well known that substantially all the world's production of titanium is produced by the Kroll process which consist of metallothermic reduction of titanium tetrachloride ($TiCl_4$/tickle) with magnesium metal in a batch retort at approximately 900 °C. This 70 year old process is technically mature with little likelihood that operating cost and capital cost can be reduced to produce titanium sponge at a significant cost reduction. Other aspects with the Kroll process is a toxic and corrosive material, tickle, requires storage and handling, the by product magnesium chloride ($MgCl_2$) must be handled, maintained anhydrous and electrolyzed to produce the reductant magnesium, and the product is a sponge which must be separated from residual $MgCl_2$ and then melted (often triple melted) and alloyed with subsequent processing into products. The Kroll processing to produce sponge results in high cost primary titanium and the down stream processing results in added cost that translates to high cost titanium products. Reduction of the cost of titanium products must include producing primary titanium at a lower cost. It has long been recognized that if titanium could be electrolytically produced the cost has the potential to be lower than Kroll produced sponge. The energy to extract titanium from its oxide is within about 10% that of aluminum from its oxide. Since aluminum is economically produced electrolytically, it should be possible to produce low cost titanium electrolytically. In the past, the plethora of investigations to electrolytically produce titanium has utilized $TiCl_4$/tickle as the feed. Since $TiCl_4$ is a covalently type bonded compound, it does not complex well with fused salts resulting in a required first reduction step to produce $TiCl_3$ or $TiCl_2$ that will complex with fused salts and when electrolyzed will deposit titanium. The multivalences of titanium are subject to frequent disproportion between the deposited titanium and the higher valences resulting in poor Columbic current experiences. Also the deposited titanium is on large cathode sheets which must be exchanged through inert gas locks complicating any production operation. These difficulties have contributed to the lack of translating an electrolytic process utilizing $TiCl_4$/tickle as a feed into commercial production.

117

DARPA recognized it was desirable to produce titanium at a cost substantially less than Kroll produced sponge or the sodium reduction of tickle which is substantially the same as magnesium reduction in processing and cost. If tickle production along with storage and handling were not necessary, cost reduction could be envisioned. Ideally, if titanium could be made directly from ore or TiO_2, cost to produce primary titanium could possible be reduced. The Fray/FFC – Cambridge process utilizes TiO_2 as a cathode [1-3] and reportably can extract oxygen and produce titanium. This was one of the processes sponsored by DARPA to produce reduced cost primary titanium. Through several years research by several investigators no pilot or commercial production process has emerged utilizing TiO_2 as a cathode to produce low oxygen content primary titanium. Uno [4-9] and others [10-11] have reported electrolytically depositing calcium that reduces TiO_2 and the product CaO is utilized to regenerate calcium in electrolysis of CaO with a graphite anode. The produced titanium has not been reported to meet the specifications equivalent to Kroll sponge and there are no reports of the process being scaled-up.

Another electrolytic process that utilizes TiO_2 as a feed to produce low cost primary titanium has been at MER which began in 2002 and received sponsorship from DARPA, as well as a SBIR jointly sponsored by the Army and DARPA in 2003.

Composite Anode Process

The program at MER that utilizes TiO_2 as a feed consists of a two step process. The first step is a carbothermic reduction of TiO_2 to a suboxide which is utilized in a composite anode with stoichiometric carbon to electrolytically produce titanium with an anode gas consisting of carbon monoxide (CO) which may also contain some carbon dioxide (CO_2). The anode gas composition is dependent on the carbothermic reduction of TiO_2 operating parameters and thus the anode composition, the fused salt composition, and the electrolysis parameters. An illustration of the overall process is shown in Figure 1 and the concept of the composite anode is illustrated in Figure 2.

The composite anode is fabricated utilizing any source of TiO_2 in the rutile or anatase structure which is mixed with carbon and heat treated to temperatures in the range of 1200-2100 °C. The anode making process can be one step or two steps which is illustrated in Figure 1. While virtually any TiO_2 – carbon combination heat treated to above 1200 °C will ultimately produce titanium when in a composite anode with stoichiometric carbon composition in electrolysis, the anode processing is governed by economics of anode production to produce low cost titanium.

During both DARPA and the Army SBIR Phase I, it was demonstrated that titanium could be electrolytically produced via the electrolysis of composite anodes. The electrolysis was performed in fused salts with liquid temperatures as low as 150 °C and as high as 1000 °C. Cursory trials were also made at above the melting point of titanium which produced titanium in a liquid state. Typical fused salt composition consisted of the alkali and alkaline earth halides as singular chlorides, fluorides or mixtures of these. It is interesting to note that in 2005 a Chinese paper [12] was published utilizing a composite anode that reported electrolytically producing high purity titanium particulate at high columbic efficiency. More recently in 2006 [13] a Chinese patent application was filed that provided more details of the processing. This Chinese published work serves to confirm the technical feasibility of the work reported here. Some typical morphologies of titanium produced at MER in the solid state is shown in Figures 3 and 4.

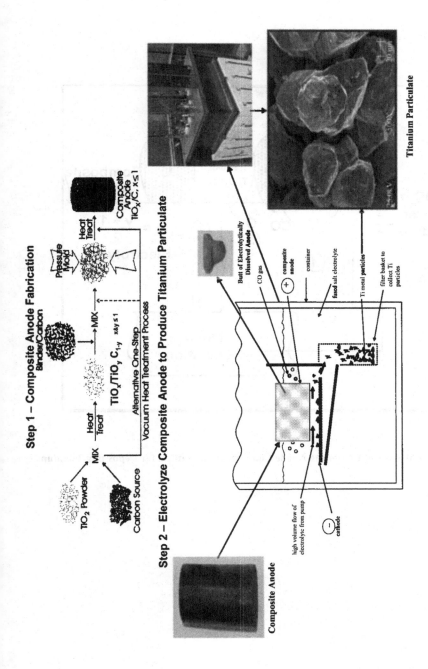

Step 1 – Composite Anode Fabrication

$TiO_x/TiO_y C_{1-y}$ $x \& y \leq 1$

TiO_2 Powder

Carbon Source

Binder/Carbon

Composite Anode $TiO_x/C, x \leq 1$

Alternative One-Step Vacuum Heat Treatment Process

Step 2 – Electrolyze Composite Anode to Produce Titanium Particulate

Composite Anode

Titanium Particulate

Butt of Electrolytically Dissolved Anode

CO gas

composite anode

container

fused salt electrolyte

Ti metal particles

filter basket to collect Ti particles

high volume flow of electrolyte from pump

cathode

Figure 1: Illustration of the Composite Anode Process to Produce Titanium Particulate from TiO_2

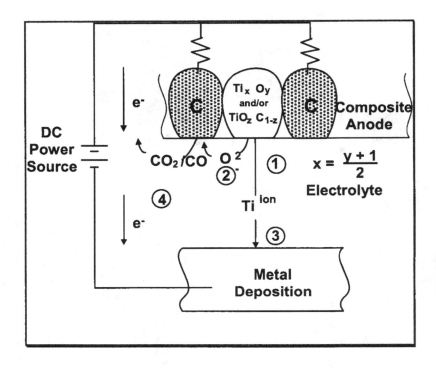

Figure 2: Schematic of Composite Anode Conceptual Mechanism for Depositing Titanium

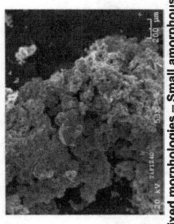

Fine Particle Size

Mixed morphologies – Small amorphous materials with crystalline material

Flakey Morphology

Good Crystalline Material

Figure 3: Variability in Morphology of Ti Particulate Produced from Electrolysis of a Composite Anode.

121

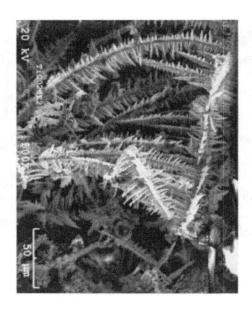

Figure 4: Occasional occurrence of dendrites deposits that maybe only a small portion of a batch deposit without high mass flow of electrolyte.

122

There is some control over the morphology of the titanium particulate produced which permits producing a product that can be used directly in powder metallurgy applications. This is a significant accomplishment as Kroll sponge must be melted and then transformed into a powder for use in powder metallurgy applications. In order for the electrolytic powder to be used directly in powder metallurgy applications the fused salt electrolyte must be removed. The removal techniques are somewhat similar to removing the $MgCl_2$ from Kroll sponge. The fused salt electrolyte is drained from the electrolytic powder and then subjected to an aqueous washing or vacuum sublimation.

Utilizing the composite anode it is possible to operate the electrolytic system in a high mass flow mode. This eliminates the need for building up titanium on large cathodes that requires air lock exchanging of the cathodes. High mass flow of the electrolyte between the composite anode and cathode sweeps the electrowon titanium particles off the cathode where they can be harvested separately and the electrowinning process can be performed on a continuous basis without opening the cell body. This concept is a significant advantage of this electrolytic composite anode process that has been further demonstrated under the DARPA and Army SBIR Phase I programs.

During Phase I, electrolytic salt residue was removed from particulate by vacuum sublimation and while hot under high vacuum the particulate was hot pressed into small cylindrical billets. This processing prevented exposure of the titanium particulate to the air until it was in a consolidated state. These small 2½" x 2½" cylindrical billets were supplied to RMI Titanium, who was a partner in part of the DARPA Phase I, and also supplied to Wright-Patterson Air Force Base (WPAFB) for characterization. WPAFB has not yet reported their results. RMI processed the billets and characterized as shown in Figure 5. Actual aerospace component parts were fabricated as shown in Figure 5 and delivered to DARPA that completed the Phase I investigation.

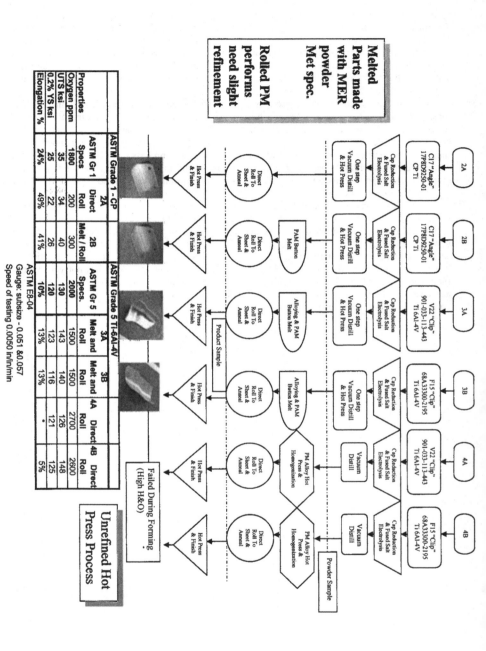

Figure 5: Phase I Titanium Parts

Properties	ASTM Gr 1 Specs	2A Direct Roll	2B Melt/Roll	ASTM Gr 5 Specs	3A Melt and Roll	3B Melt and Roll	4A Direct Roll	4B Direct Roll
Oxygen ppm	1800	200	300	2000	1500	1500	2700	2600
UTS ksi	35	34	40	130	143	140	126	148
0.2% YS ksi	25	22	26	120	123	116	121	125
Elongation %	24%	49%	41%	10%	13%	13%	·	5%

ASTM E8-04
Gauge: subsize - 0.051 &0.057
Speed of testing 0.0050 in/in/min

Based on these feasibility demonstrations, DARPA has awarded a Phase II program to a consortium including DuPont which will scale-up the process to 500 lbs/day. The Phase II began on August 17, 2006. A commercial demonstration cell begins operation in October 2006 whose operation will provide design data for the 500 lbs/day system. A 3000 lbs/day cell system is being considered for the next phase.

References

1. G. Z. Chen, D. J. Fray, and T. W. Farthing, Nature, 407 (2000), 361-364.
2. G. Z. Chen, D. J. Fray, *Electro-Deoxidation of Metal Oxides*, Light Metals, TMS 2001.
3. C. Schwandt and D. L. Fray, *Determination of the Kinetics Pathway in the Electrochemical Reduction of Titanium Dioxide in Molten Calcium Chloride*, Electroctimica Acta 51, 2005, 66-76.
4. T. H. Okabe, T. Oishi, and K. Ono, *Preparation and Characterization of Extra-Low-Oxygen Titanium*, Journal of Alloys and Compounds, 184, (1992), 43-56, JALCOM 160.
5. T. H. Okabe, M. Nakamura, T. Oishi, and K. Ono, *Electrochemical Deoxidation of Titanium*, Metallurgical Transactions B, Volume 24B, June 1993, 449.
6. Ryosuke, O. Suzuki, Masayuki Aizawa, Katsutoshi Ono, *Calcium-Deoxidation of Niobium and Titanium in Ca-Saturated $CaCl_2$ Molten Salt*, Journal of Alloys and Compounds, 228, (1999), 173-182.
7. Katsutoshi Ono and Ryosuke O. Suzuki, *A New Concept for Producing Ti Sponge: Calciothermic Reduction*, JOM, February 2002.
8. Ryosuke O. Suzuki and Katsutoshi Ono, *A New Concept of Sponge Titanium Production by Calciothermic Reduction of Titanium Oxide in the Molten $CaCl_2$*, Electrochemical Society Proceedings, Volume 2002-19.
9. Ryosuke O. Suzuki, Koh Teranuma, and Katsutohi Ono, *Calciothermic Reduction of Titanium Oxide and In-Situ Electrolysis in Molten $CaCl_2$*, Metallurgical and Materials Transitions B, Volume 34B, June 2003-287.
10. Ryosuke O. Suzuki and Shuichi Inoue, *Calciothermic Reduction of Titanium Oxide in Molten $CaCl_2$*, Metallurgical and Materials Transaction B, Volume 34B, June 2003-277.
11. Ryosuke O. Suzuki, *Calciothermic Reduction of TiO_2 and In-Situ Electrolysis of CaO in the Molten $CaCl_2$*, Journal of Physics and Chemistry of Solids, 66, (2005), 461-465.
12. S. Jiao, X. Hu, H. Zhu, *Titanium Electrolysis using TiC_xO_y Anode Prepared Through Carbothermic Reduction of Titanium Dioxide*, 7th International Symposium on Molten Salts Chemistry and Technology, Volume II, p. 867, August 29-September 2, 2005, Toulouse, France.
13. S. Jiao, X. Hu, H. Zhu, *Method of Preparation of Pure Titanium using Anode Electrolysis of Titanium (I) Oxide/Titanium Carbide Solid Solution*, International Patent Application No. 2005 10011684.6, December 28, 2005.

A FUNDAMENTAL INVESTIGATION ON RECOVERY OF TITANIUM FROM TITANIUM-BEARING BLAST FURNACE SLAG

Tao Jiang, Haigang Dong, Yufeng Guo, Guanghui Li, Yongbin Yang

School of Minerals Processing & Bioengineering; Central South University, Changsha, Hunan, 410083, China

Keywords: Ti-bearing blast furnace slag, Sulfuric acid leaching, Activation

Abstract

Based on physical and chemical characteristics of Ti-bearing blast furnace slag, H_2SO_4 leaching is investigated to recover titanium from Ti-bearing blast furnace slag. It is shown that the 72.26% of TiO_2 leaching can be obtained under the conditions of H_2SO_4 concentration of 50%, reaction temperature of 100□, reaction time of 1h, liquid to solid ratio of 10 and particle size of -0.5mm.The leaching mechanism is investigated from thermodynamics, mineral crystallization and morphology of slag.

Introduction

There are lots of V-Ti bearing magnetite ores in Pan-Xi area of China. Reserves of titanium in the area account for 90.54% of total titanium resource. According to present process of Pan Steel, a large amount of blast furnace slag bearing 22%-25% TiO_2 has been produced. Nowadays, this slag is piled largely as waste materials [1-7].

To recover TiO_2 from the titanium-bearing blast furnace slag (TBFS) effectively, many studies have been carried out. The investigations by Qifu CHEN [8] show that the TiO_2 extraction of 88% can be obtained by concentrated sulfuric acid (H_2SO_4 mass fraction 89%~93%) from the slag. After removing iron-bearing minerals by magnetic separation, TBFS is leached under pressure with dilute H_2SO_4 (20%~60% mass fraction), and leaching of TiO_2 is over 85% [9]. But additional consumption of sulfuric acid on dissolving CaO, MgO and Al_2O_3 in matrix may cause higher cost and more environmental contamination. Otherwise, leaching under pressure needs air-tight and pressure equipment, which will lead to more cost and complex process. Thereby, there are no effective methods to recover TiO_2 from TBFS.

In this paper, the feasibility is investigated on recovering of TiO_2 from TBFS by leaching with dilute H_2SO_4 in atmosphere, and the mechanism is discussed at the same time.

Materials and Methods

Characteristics of Materials

The main material used in the studies is supplied by Panzhihua Iron and Steel Company of China, and the sample is a kind of water quenched slag. The chemical compositions of tested sample are shown in Table I. The slag contains TiO_2 22.36%; the total content of CaO, SiO_2, MgO and Al_2O_3 is over 70%.

Table I Main Chemical compositions of TBFS

Compositions	TFe	TiO_2	CaO	SiO_2	MgO	Al_2O_3	Na_2O	K_2O	MnO
Contents/%	1.16	22.36	27.81	23.46	7.98	13.89	0.35	0.73	0.73

The size distribution of tested sample is shown in Table II. It is shown that, the slag is cooled rapidly when treated with water directly, so its granularity is fine relatively, the content of -3.5mm is over 95%.

Table II Size Distribution of TBFS sample

Granularity/mm	+3.5	3.5~2	2~1	1~0.5	-0.5
Percentage composition/%	4.84	24.00	21.52	24.64	25.00

Methods

At the beginning of each trail, TBFS sample is added into the test container, and blends with H_2SO_4 liquor (the concentration of H_2SO_4 liquor is prepared). The mixture is agitated and leached isothermally at uniform temperature for a given period. After filtering, drying, weighing, the content of TiO_2 in the residue is analyzed and the leaching of TiO_2 is calculated by following equation:

$$\eta = (1 - \frac{\beta \bullet \gamma}{100 \, \alpha}) \times 100$$

η: TiO_2 leaching, %
α: TiO_2 grade of raw material, %
β: TiO_2 grade of the filter residue, %
γ: Productivity of the filter residue, %

Results

The main Ti-bearing mineral in the slag is perovskite ($CaTiO_3$). During the H_2SO_4 leaching process following reaction will take place [9]:

$$CaTiO_3 + 2H_2SO_4 = TiOSO_4 + CaSO_4 + 2H_2O$$

Factors affecting leaching results involve H_2SO_4 concentration in liquor, leaching time, leaching temperature, particle size, ratio of liquid to solid (L/S), stirring speed, and so on.

Effect of H_2SO_4 Concentration

Effect of H_2SO_4 concentration (mass fraction) on TiO_2 leaching is shown in Fig.1. The slag is leached under the conditions of leaching temperature 80°C, leaching time 2 h, L/S ratio 10, particle size -3.5 mm and stirring speed 400 r/min.

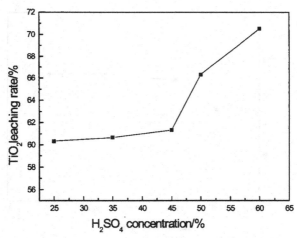

Fig.1 Effect of H_2SO_4 concentration on TiO_2 leaching

It is shown in Fig.1 that, the leaching of TiO_2 increases gradually with H_2SO_4 concentration. When H_2SO_4 concentration reaches 50%, TiO_2 leaching of 66.32% can be obtained. However, a white colloid substance is observed during leaching process. Its compositions are obtained by chemical analysis as shown in Table III.

Table III The chemical composition of white colloid substance

Elements	O	S	Ca	Ti	Al	Si	Fe
Percentage/%	56.40	17.5	23.79	1.77	0.06	0.33	0.04

It can be found that, from Table III, the main elements are O, S and Ca, in terms of whose mol ratio in white colloid substance, we can infer that the substance is $CaSO_4$ with crystal water. The white colloid substance hinders the diffusion of H_2SO_4 on surface of slag. So the leaching of TiO_2 is affected. With increase of H_2SO_4 concentration, viscosity of ore pulp is decreased and the leaching of TiO_2 is increased.

Effect of Leaching Time

Under the leaching conditions of H_2SO_4 concentration 50%, temperature 80°C, L/S ratio 10, particle size -3.5 mm and stirring speed 400 r/min, the effect of leaching time on TiO_2 leaching is shown in Fig.2.

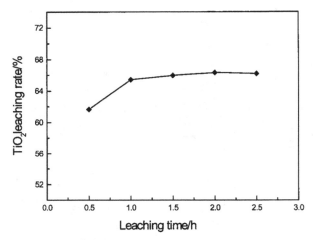

Fig.2 Effect of leaching time on TiO$_2$ leaching

It can be seen from Fig.2 that TiO$_2$ leaching increases with increase of leaching time, but when the time is over 1 h, the TiO$_2$ leaching maintained constant almost. Meanwhile, it can be found that the quantity of white colloid substance increased gradually with leaching time. TiO$_2$ leaching is affected.

Effect of Leaching Temperature

Under the conditions of H$_2$SO$_4$ concentration 50%, leaching time 1 h, L/S=10, particle size-3.5 mm and stirring speed 400 r/min, effect of leaching temperature on TiO$_2$ leaching is shown in Fig.6.

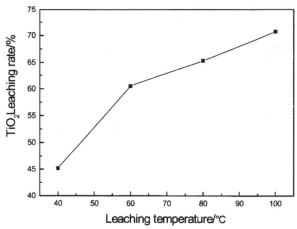

Fig.3 Effect of leaching temperature on TiO$_2$ leaching

From Fig.3, it can be concluded that TiO_2 leaching is affected by temperature obviously; increasing temperature can be favorable for TiO_2 leaching. And TiO_2 leaching is up to 70.74% when leaching temperature is 100℃. Also, it can be found that the quantity of white colloid substance decreased with increase of leaching temperature, which is in favor of diffusion of H_2SO_4 on the surface of mineral and improves TiO_2 leaching.

Effect of Particle Size

Effect of particle size on TiO_2 leaching is carried out under the conditions, which were H_2SO_4 concentration 50%, leaching time 1 h, L/S=10, leaching temperature 100℃ and stirring speed 400 r/min. The results are shown in Fig.7.

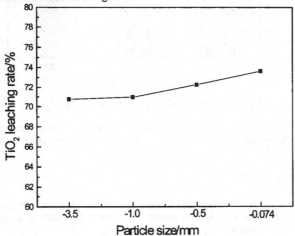

Fig.4 Effect of particle size on TiO_2 leaching

It is shown in Fig.4 that the finer particle size of material is, the more TiO_2 leaching increases. The leaching of TiO_2 is 72.26% when particle size is less than 0.5mm, while TiO_2 leaching is only 73.62% when particle size of raw material is less than 0.074mm. The leaching process is an out-of-phase reaction between solid and liquid and the reaction is put up on the surface of slag. According to theory of shrinking unreacted core model, decrease of particle size can make thickness of interface liquid film on particles reduced and ion rate passing the diffusion film increase. However, particle size is too fine, instant reaction is quick, which will lead to coherence of mineral particles on reactor surface. Suitable particle size is less than 0.5mm.

Effect of Ratio of Liquid to Solid (L/S)

Effect of L/S on TiO_2 leaching is carried out under the conditions of 50% H_2SO_4 concentration, 1 h leaching time, -0.5mm particle size, 100℃leaching temperature and 400 r/min stirring speed. The results are shown in Fig.5.

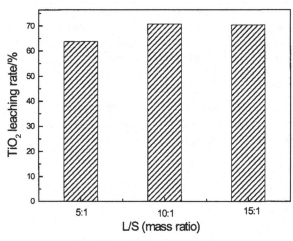

Fig.5 Effect of L/S on TiO_2 leaching

The TBFS is loose and bulky, so ratio of liquid to solid can not be too low, otherwise, liquid will not cover with solid completely. From Fig.5, L/S has no obvious effect on TiO_2 leaching when L/S ratio is over 10.

Effect of Stirring Speed

Effect of stirring speed on TiO_2 leaching is carried out under reaction conditions 50% H_2SO_4 concentration, 1 h leaching time, -0.5mm particle size, and 100□ leaching temperature and 10 L /S ratio. The results are shown in Fig.6.

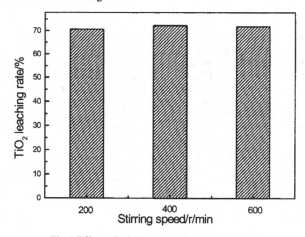

Fig.6 Effect of stirring speed on TiO_2 leaching

Fig.6 shows that TiO_2 leaching does not change a lot with increase of stirring speed. Also, when stirring speed is over 200r/min, reaction is not affected at all. It can be concluded that the reaction is controlled by chemical reaction but diffusion.

Mechanism of Leaching Process

Thermodynamics

Reaction involved in sulfuric acid leaching of TBFS is very complex. It is very necessary to find out possibility and maximum extent of reaction. Here, thermo-dynamics involved reactions are calculated, and the possibility and conditions are analyzed and discussed.

Standard Gibbs free energy (ΔG_T^θ) forTiO^{2+} and Ti^{3+} can not be obtained by the following equations [10-12]:

$$TiO^{2+} + H^+ + 4e \rightarrow Ti + H_2O \qquad \varphi^o = -0.89V \qquad (3)$$

$$\Delta_r G_{298}^o = -n\varphi^o F = 343.54 kJ\,/\,mol \qquad (4)$$

$$TiO^{2+} + H^+ + e \rightarrow Ti^{3+} + H_2O \qquad \varphi^o = 0.1V \qquad (5)$$

From Eq.1 to Eq.2, the following results can be known:

$$\Delta G_{f,298}^o (TiO^{2+}) = -5890.73 kJ\,/\,mol \qquad (6)$$

$$\Delta G_{f,298}^o (Ti^{3+}) = -353.19 kJ\,/\,mol \qquad (7)$$

Based on the results, by consulting manual book of thermodynamic data, $\Delta_r G_{298}^o$ of possible reactions between H_2SO_4 and TBFS are calculated and given in Table IV.

Table IV $\Delta_r G_{298}^o$ of possible reactions between H_2SO_4 and TBFS

Equations	$\Delta_r G_{298}^o$ /kJ·mol^{-1}
$CaTiO_3 + 4H^+ + SO_4^{2-} \rightarrow CaSO_4 + TiO^{2+} + 2H_2O$	-91.12
$Ti_3O_5 + 8H^+ \rightarrow TiO^{2+} + Ti^{3+} + 4H_2O$	261.84
$CaO \cdot TiO_2 \cdot SiO_2 + 4H^+ + SO_4^{2-} \rightarrow TiO^{2+} + CaSO_4 + 2H_2O$	5.51
$TiO_2 + 2H^+ \rightarrow TiO^{2+} + H_2O$	141.82
$CaO + 2H^+ + SO_4^{2-} \rightarrow CaSO_4 + H_2O$	-314.51
$MgO + 2H^+ \rightarrow Mg^{2+} + H_2O$	-87.76
$Al_2O_3 + 6H^+ \rightarrow 2Al^{3+} + 3H_2O$	-91.58

It is shown in Table IV that only the $\Delta_r G_{298}^o$ value of reaction between $CaTiO_3$ and H_2SO_4 is negative, which means only the reaction between $CaTiO_3$ and H_2SO_4 can take place.

The main crystal mineral phase is $CaTiO_3$ and other Ti-bearing minerals occur in non-crystal substance phase. This partly explains why the leaching of TiO_2 is 70% or more. At the same time, the quantity of CaO dissolved by H_2SO_4 is great. And white colloid substance ($CaSO_4 \cdot nH_2O$) is produced, which baffles more increase of TiO_2 leaching.

Effect of crystallization

Because long-range disorder of non-crystalline substance makes its inner energy not stay at

133

lowest state. Non-crystal substance lies in metastable state [13-14].

Main minerals in TBFS sample are analyzed by X-Ray Diffraction (XRD).The result is given in Fig.7. XRD results show that there is one crystal mineral as perovskite ($CaTiO_3$) in the sample only, but a big and tremendous diffraction peak centered at about $30°\ 2\theta$ indicates that there exists lots of non-crystalline substance in the sample. Therefore, according to XRD results, titanium occurs as $CaTiO_3$ mainly in the slag, but there should be a fraction of titanium existing in the non-crystalline substance.

In addition, formation and growth of crystal core are limited by water quenching, which will cause crystal defect of $CaTiO_3$ and formation of amorphous substances. It can be concluded that water quenching is an activation process. Ti-bearing components that have strong activation in non-crystal substance can be dissolved by H_2SO_4. It is favorable to extract titanium by hydrometallurgical process.

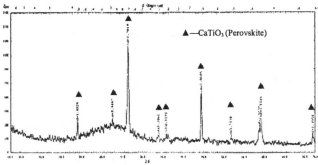

▲ —$CaTiO_3$ (Perovskite)

Fig.7 XRD results of TBFS sample

Effect of Morphology of Slag

Macroscopic appearance and microcosmic morphology of TBFS sample are recorded by digital camera and scanning electronic microscope (SEM). The results are given in Fig.8 and Fig.9 respectively.

Fig.8 Macroscopic appearance of TBFS sample

134

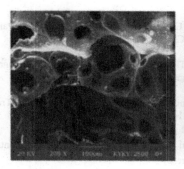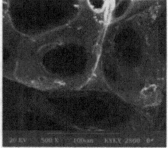

Fig.9 Microcosmic morphology of TBFS sample recorded by SEM

From Fig.8 and Fig.9, it can be concluded that the slag is a loose and porous powder, there exists a large number of microporosity in its structure. H_2SO_4 easily penetrated through inner microporosity of slag, which increases the reaction and reduces reaction time.

Conclusions

Technologies and fundamental of H_2SO_4 leaching to recover titanium from Ti-bearing blast furnace slag are investigated in this paper.

1) Leaching experiments show that 72.26% TiO_2 leaching can be leached out under the conditions of 50% H_2SO_4 concentration, 1 h leaching time, -0.5mm particle size, 100°C leaching temperature, and 10 ratio liquid to solid. However, formation of white colloid substance ($CaSO_4 \cdot nH_2O$) baffled more increase of TiO_2 leaching.

2) The leaching mechanism is investigated from thermodynamic, crystallization and morphology of slag. It is shown that in this water quenched slag; $CaTiO_3$ is only Ti-bearing component which can react with H_2SO_4. The slag is activated by water quenching and lots of non-crystalline substance and $CaTiO_3$ with crystal defect which are well-reaction activation are formed, which are favorable for leaching TiO_2. At last, TBFS is a loose and porous powder; there exists a large number of microporosity in its structure, which increases the reaction and reduced reaction time.

References

1. Hegui DU. "Theory for Blast furnace refining V and Ti-bearing magnetite", (Beijing: Metallurgy Industry press, 1996.) (In Chinese)

2. Wei MO. "Titanium metallurgy", (Beijing: Metallurgy Industry press, 1998) (In Chinese)

3. Kang SUN. "Physical and Chemistry of Extracting Metallurgy of Titanium", (Beijing: Metallurgy Industry press, 2001.) (In Chinese)

4. Xi-qing WANG. "Blast furnace refining V and Ti-bearing magnetite", (Beijing: Metallurgy Industry press, 1994.) (In Chinese)

5. Liping WANG et al. "Distribution and production status of Titanium resources in China", Chinese Journal of Rare Metals, 8(1) (2004), 265-267. (In Chinese)

6. Ke WANG, Huijuan MA. "Comprehensive Utilization of Titanium resource of Panzhihua", Vanadium and Titanium, (6) (1992), 10-19. (In Chinese)

7. Qifu CHEN. "Extended test of extracting TiO_2 and Sc_2O_3 from Pan Steel Ti-bearing blast furnace slag", Iron, Steel, Vanadium and Titanium, 16(3) (1995), 64-68. (In Chinese)

8. Xiaohua LIU, Zhitong SUI. "Pressure leaching of Ti-bearing blast furnace slag", (Transaction of nonferrous metal of China, 12(6) (2002), 1281-1284. (In Chinese)

9. Zhuqing GONG. "Introduction of Theoretical Electro-chemistry", (Changsha : Central South University of Technology press, 1988.) (In Chinese)

10. Qingheng ZENG. "Physical Chemistry", (Changsha: Central South University of Technology press, 1996.) (In Chinese)

11. Kathryn C. Sole. "Recovery of titanium from the leach liquors of titaniferous magnetites by solvent extraction Part 1. Review of the literature and aqueous thermo- dynamics", Hydrometallurgy, 51(1999), 239-253.

12. Dalun YE, Jianhua HU. "Handbook of Thermodynamic Data on Practical inorganic matter", (Beijing: Metallurgy Industry press, 2002.) (In Chinese)

13. Bingchu YANG, Xingang ZHONG. "Solid Physics", (Changsha: Central South University press, 2002.) (In Chinese)

14. Shengtao HUANG. "Structure and Structural analysis of non-crystal material", (Beijing: Science press, 1987.) (In Chinese)

EFFECT OF Al₂O₃ CONTENT ON THE ELECTRO-WINNING OF Ti IN DC-ESR OPERATION

M. Kawakami[1], T. Takenaka[1], T. Kawabata[2], A. Matsuyama[2] and S. Yokoyama[1]
[1]Department of Production Systems Engineering,
Toyohashi University of Technology
Tempakucho-aza-Hibarigaoka 1-1, Toyohashi 441-8580, Japan
[2]Graduate course, Toyohashi University of Technology

Keywords: Electro-winning of Ti-Al, Molten slag electrolysis, DC-ESR, Reaction mechanism

Abstract

In order to make titanium ingot directly from molten slag, the metal pool was tried to form and used as the cathode. The electro-slag remelting (ESR) unit was used for the purpose. The unit is composed of water-cooled copper mold and the DC power source of 2000 A and 100V. The graphite electrode was used for anode and the base plate was used for cathode. The slag was composed of the CaF₂-CaO-Al₂O₃ system. By the addition of Al₂O₃ to the slag, the electro-deposit of Ti-Al was obtained. The current efficiency increased with the supplied electric power. It is concluded that the mechanism of high current efficiency should be the formation of liquid metal deposit due to high heat generation and the suppression of metal fog formation and sub-oxide formation of titanium.

Introduction

Titanium has such good properties as high specific strength and high corrosion resistance. The availability of titanium has been limited because of its high production cost. The titanium metal ingot is now produced by the Kroll process which is schematically shown in Fig. 1. Namely, the purified TiO₂ is chlorinated to TiCl₄ which is produced by magnesium to be titanium metal in sponge form. The sponge titanium is distilled to eliminate the residual magnesium and magnesium chloride. Then, the sponge titanium is re-melted in vacuum several times to get the ingot form. Recently, the molten salt electrolysis of titanium has been proposed[1,2]. But, in their process, the electro-deposit is in solid state, resulting in the subsequent re-melting process. If the solid ingot of titanium can be obtained by direct electrolysis from molten slag, it might be of some help to reduce the production cost. In general solid electrodeposits from molten slag are likely to be dendritic, which necessitates such a succeeding process as vacuum arc re-melting. If the electro-deposit can be obtained in liquid state, the above problem will be solved. The proposed process is schematically shown also in Fig. 1. However, it is considered very difficult to hold liquid titanium by the ordinal means because of high melting point and high reactivity.

In the previous work, it was shown that the electro-winning of titanium was possible with the aid of a small scale ESR unit[3-5]. But the maximum cathodic current efficiency was only 18 % by using CaO-CaF2 binary slag. One of the reasons was that the metal pool could not be formed because of the shortage of electric power supply. It was shown that the metal pool of titanium could not be formed in the ESR operation[6]. Then, the carbon steel plate was used as the base plate cathode. But, even so, the metal pool could not be obtained using binary CaO-CaF₂ slag as will be shown latter. In order to obtain the metal pool, more heat should be supplied. On way is to increase electrical power by increasing current and voltage. But the voltage could not be raised, even if the current was raised. Thus, in the present work, some amount of alumina was added to the slag in order increase the resistivity of the slag[7]. The effect was examined from kinetic viewpoint.

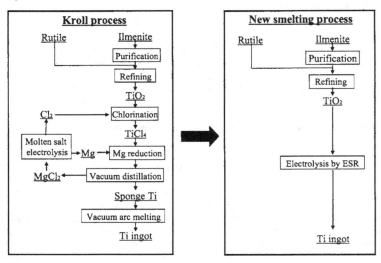

Fig. 1 Schematic diagram of Kroll process and the proposed process.

Principle

If an ESR apparatus is operated in a DC reverse polarity mode, cathodic reactions should occur at the metal pool on the bottom of cell. Since the process should proceed at as high temperature as 2000 K, the titanium source should be TiO_2, rather than chlorides because of high vapor pressure. If the slag contains TiO_2, the virtual cathodic reaction is expected to be

$$Ti^{4+} + 4e = Ti \qquad (1)$$

A graphite rod suspended in the middle of cell is used as an anode. The virtual anodic reaction is expected to be

$$O^{2-} + C = CO + 2e \qquad (2)$$

Then, the overall reaction is expected to be

$$TiO_2 + 2C = Ti + 2CO \qquad (3)$$

Therefore, the bath should have enough solubility of TiO_2 at the concerned temperature and its main components should be more stable electrochemically than TiO_2. In the present work, CaF_2-CaO-Al_2O_3 was selected as an electrolytic bath. I this case, the electro-deposition of aluminum is also expected. The mechanism of cathodic reaction will be discussed.

138

Experiments

Apparatus

The experimental apparatus is schematically shown in Fig.2. The inner diameter of the vessel was 110 mmφ and the height was 400 mm. The side-molds and base-molds were cooled by water stream. The water cooling system was equipped with a flow meter and two thermometers at the inlet and outlet of the molds. The heat loss from the molds can be estimated from water flow rate and the temperature change.

A graphite rod was used as the anode. The diameter of the rod was 50 mmφ and the length was c.a. 1 m. The rod was machined so that the cross section was of star like shape and the top was conical. The rod was driven up and down with the aid of an electrode driving unit. The carbon steel base-plate of 105 mmφ and 20 mm thickness was put on the base-mold and was used as the cathode.

The maximum capacity of power unit was 200kVA. The maximum DC current and voltage were, respectively 2.0 kA and 100 V. The electric power unit was of constant current type and equipped with voltage and current recorders. The positive terminal of the power unit was connected to the graphite anode, and the negative terminal was to the base-mold. The side-molds were electrically insulated from the base-mold so that the current is expected to flow perfectly from the graphite rod to the base-plate.

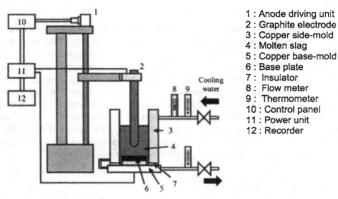

1 : Anode driving unit
2 : Graphite electrode
3 : Copper side-mold
4 : Molten slag
5 : Copper base-mold
6 : Base plate
7 : Insulator
8 : Flow meter
9 : Thermometer
10 : Control panel
11 : Power unit
12 : Recorder

Fig. 2 Schematic diagram of experimental apparatus.

Slags

The slag was synthesized by ourselves. The CaF_2-CaO-Al_2O_3 mixture was pre-melted in the graphite crucible and quenched on the water cooled copper plate. The solidified mixture was crushed and kept in the dry box at about 450 K. The composition of slag is shown in Table 1. The TiO_2 powder was added to the molten mixture after the melt down of the mixture so that the TiO_2 content should be about 10 mass%.

Experimental process

The electrode was driven down so that the tip should just touch the base-plate. About 1 kg of the CaO-CaF_2 mixture was supplied to the vessel. When the power unit was switched on, the electric arc was generated initially. By the arc, the primary slag powder melted down gradually. After it melted down completely, the slag powder was supplemented. The arc stopped and the current started to flow through the molten slag.   The total amount of slag was 2.7 kg.   Three hundred grams of TiO$_2$ powder was added to the melt.

After stopping the power supply, the electrode was pulled up.   The weight changes of the base-plate and the graphite electrode were measured.   In the case where some granular deposits were observed in the slag, the solidified slag was crashed up to collect the metallic pieces.   The amount of them was also measured.   The pictures of base-plate surface and solidified slag were taken by a digital camera.   The electro-deposit was examined by SEM and EPMA.

Table 1   Slag composition (wt%)

Composition [wt%]	CaF$_2$	CaO	Al$_2$O$_3$
base	80	20	
1	70	20	10
2	60	20	20
3	50	20	30

Results

Process outline

Figure 3 shows the change in current and voltage with time when the slag composition was CaF$_2$-20%CaO.   The current was increased step wise up to 1, 500 A.   But the voltage did not increased so much, showing that the high power could not be obtained with this binary slag. Figure 4 shows the photo of the surface of base-plate after the experiment.   A very small portion appeared to be melted.   Thus, the sufficient amount of metal pool was not formed in this experimental run.   Figure 5 shows the change in current and voltage with time when the slag composition was CaF$_2$-20%CaO-30%Al$_2$O$_3$.   The current was controlled constant at 1,200A. At first, the voltage was about 40V and later jumped up to about 80V.   Thus, much higher electrical power was supplied than in Fig. 3.   Figure 6 shows the photo of the surface of base-plate after the experiment.   A large portion appeared to be melted.   The electro-deposit can be seen on the base plate.   In this run, the amount of deposit was about 100g.

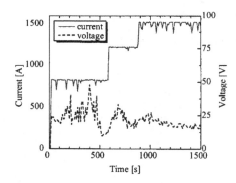

Fig. 3 Change in current and voltage with time: Fig. 4 Photo of the surface of base-plate:
    CaF$_2$-20%CaO.   CaF$_2$-20%CaO.

140

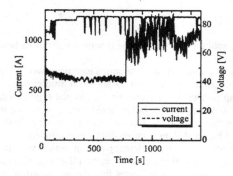

Fig. 5 Change in current and voltage with time: CaF₂-20%CaO-30%Al₂O₃.

Fig. 6 Photo of the surface of base-plate: CaF₂-20%CaO-30%Al₂O₃.

Electro-deposit

Figure7 shows the electro-deposit corrected in the slag phase in the run where the photo of the surface of base-plate is shown in Fig.6.   It seems that the electro-deposit was melted during the operation.   The composition of it was analyzed by EPMA and shown in Table 2 together with the result of another run.   The electro-deposit contained about 20% of Al and other minor impurities such as Fe, Si and Ca.   The iron may come from the base-plate.   The silicon may come from the contamination of slag, because the silica tube was used during melting of slag. The calcium may come from the slag entrainment.   The aluminum should be the reduction product of Al_2O_3 by the electrolysis. The formation mechanism will be discussed later.   Thus, by the present method, the Ti-Al alloy was eventually obtained rather than pure Ti.   From the binary phase diagram of Ti-Al, it can be seen that the formed alloy might be Ti_3Al.

Table 2 Composition of electro-deposit.

CaF₂-CaO-Al₂O₃	Ti	Al	Fe	Ca	Si
60-20-20	bal.	21	—	1.7	4.2
50-20-30	bal.	23	3.1	—	4.3

Fig. 7 Photo of electro-deposit corrected
　　　　in the slag phase

Current efficiency

Since the electro-deposit was mainly composed of Ti and Al, the current efficiency for cathodic reaction was calculated by the following equation.

$$\eta_{CATH} = \{\{(4FW_{Ti}/(M_{Ti})+(3FW_{Al}/M_{Al})\}/Q\} \times 100 \ (\%) \quad (4)$$

where, F is Faraday constant, W_{Ti} is the weight of titanium, M_{Ti} is the atomic weight of Ti, W_{Al} is the weight of Al, M_{Al} is the atomic weight of Al and Q is the total electricity supplied.   The weight of Ti and Al were obtained from the total weight and the composition of electro-deposit.

141

The current efficiency for anodic reaction was calculated by the following equation.

$$\eta_{ANO} = \{2FW_C/(M_C Q)\} \times 100 \ (\%) \tag{5}$$

where, W_C and M_C are weight losses of the graphite electrode and the atomic weight of carbon, respectively.

Figure 8 shows the cathodic current efficiency as a function of the supplied electric power. The efficiency increased with the electric power. The data seemed to lay on a smooth curve, regardless of the slag composition. The slag temperature should increase by the high power supply. This enabled that the electro-deposit was not dendritic but in liquid state. As shown in Fig.7, the electro-deposit was in liquid state in this run. If the electro-deposit is in liquid state, the cathodic current efficiency might be high. Figure 9 shows the anodic current efficiency as a function of the supplied electric power. The anodic current efficiency also increased with the electric power. The efficiency of less than 100% shows that the other reaction than equation (2) occurs at the anode. The efficiency of larger than 100% indicates that the air oxidation of graphite occurred simultaneously with the electrode reaction, because the experiment was performed in the ambient atmosphere.

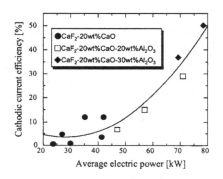

Fig. 8 Cathodic current efficiency.

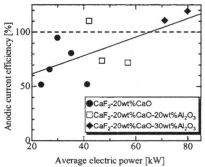

Fig. 9 Anodic current efficiency.

Reaction mechanism

In CaO-CaF$_2$-TiO$_2$ slag, the reaction of Eq. (1) is expected. However, it is conceivable that the dominant cathodic reaction should be the reduction of Ca^{2+}, because the dominant cation species is Ca^{2+}. Namely,

$$Ca^{2+} + 2e = Ca \tag{6}$$

This metallic calcium can reduce Ti^{4+} ion.

$$Ti^{4+} + 2Ca = Ti + 2Ca^{2+} \tag{7}$$

The sequence of Eqs. (6) and (7) is equal to Eq. (1).

In this slag system, the cathodic current efficiency was 12% at maximum and the anodic current efficiency did not exceed 100%. These low current efficiency at cathode and anode indicate that the current was not completely transfer by the reactions (1) and (2).

One of the mechanisms of this low current efficiency might be the formation of metal fog as shown in Fig. 10. Namely, Ca^{2+} ion is reduced to metallic Ca. But the reduced Ca forms the so-called metal fog, because the CaO-CaF$_2$ system has some solubility of metallic Ca[8]. Then, the Ca in the form of metal fog will migrate to anode where it is oxidized again to Ca^{2+} ion. This mechanism might operate, when the electrode distance is small.

Another mechanism is shown in Fig. 11. The metallic Ti formed by reaction (1) might react with Ti^{4+} to form Ti^{2+} ion by the reaction (8). The Ti^{2+} ion might migrate to anode and reoxidized again to Ti^{4+}. Thus, the current is transferred by the migration and reoxidation of titanium ion.

$$Ti + Ti^{4+} = 2Ti^{2+} \tag{8}$$

This mechanism might operate, when the distance and electrode is large. In this case, the heat transfer to the cathode at which the reaction should be endothermic, should be suppressed. The electro-deposit of Ti might be dendritic so that the reaction (8) should be accelerated.

In CaO-CaF$_2$-Al$_2$O$_3$-TiO$_2$ slag, the Al^{3+} cation can be elecrolytically reduced or reduced by the metallic Ca which is formed by Eq. (6) as shown in Fig. 12.

$$Al^{3+} + 3e = Al \tag{9}$$

$$Al^{3+} + 3/2Ca = Al + 3/2Ca^{3+} \tag{10}$$

The formed aluminum has no solubility to the slag and thus does not form the metal fog. At this high temperature, the aluminum might precipitate in liquid form. The aluminum can also reduce Ti^{4+} cation.

$$Ti^{4+} + 4/3Al = Ti + 4/3Al^{3+} \tag{11}$$

The metallic titanium can be formed in three ways, namely by Eqs. (1), (7) and (11). But, it might be absorbed in liquid aluminum to form Ti-Al alloy as shown in Fig. 12. By the reactions (9) to (11), the metal fog formation and sub-oxide formation by Eq. (8) might be suppressed. This might be the reason for the high current efficiency.

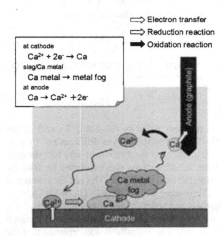

Fig. 10 Schematic diagram of metal fog. formation.

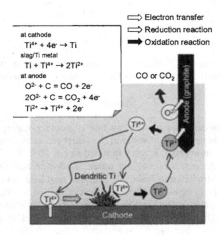

Fig. 11 Schematic diagram of titanium sub-oxide ions migration.

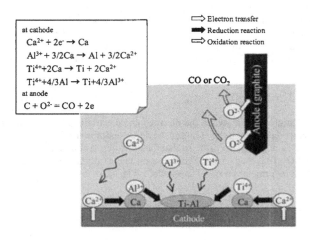

at cathode

$Ca^{2+} + 2e^- \rightarrow Ca$

$Al^{3+} + 3/2Ca \rightarrow Al + 3/2Ca^{2+}$

$Ti^{4+} + 2Ca \rightarrow Ti + 2Ca^{2+}$

$Ti^{4+} + 4/3Al \rightarrow Ti + 4/3Al^{3+}$

at anode

$C + O^{2-} = CO + 2e$

CO or CO$_2$

⇨ Electron transfer

➡ Reduction reaction

⇨ Oxidation reaction

Fig. 12   Shematic diagram of Ti-Al alloy formation.

Heat balance

The heat balance of the process must be considered carefully to form a large metal pool. The enthalpy change of reaction (3) was calculated at 1973 K as ΔH = +710.6 kJ/mol.[9] If the current induces the reaction perfectly, the necessary heat to induce reaction (3) is given by

$$Q_{(3)} = \Delta H \cdot I / 4F \qquad (12)$$

where, I is current and F is Faraday constant.   When, I = 800A, then, $Q_{(3)}$ = 1.47 kJ/s. The Joule heat generation in the slag can be roughly estimated in the alternative way.

$$Q_{slag} = I_{av}E_{av} \qquad (13)$$

where, I_{av} and E_{av} are average current ant average voltage.   Using equation (13), Q_{slag} was estimated as c.a. 64kJ/s.   The heat loss from molds was estimated by

$$Q_{LOSSW} = R_W C_p (T_O - T_I) \qquad (14)$$

where, R_W is the flow rate of cooling water, C_p is the heat capacity of water, T_O and T_I are temperature of water at the outlet and inlet, respectively.   The heat loss was different run by run. On average, it was estimated as c.a. 20 kJ/s.   Thus, c.a. 26 % of the Joule heat was lost through the molds by cooling water.   Radiation heat loss from a bath surface was estimated as 10% by Mitchell and Joshi.[10]   The total heat loss can be estimated as c.a. 25 k./s.   The net heat which is Q_{slag} minus total heat loss was estimated as 39kJ/s.   Comparing this value with $Q_{(3)}$ = 1.47 kJ/s, it can be seen that the present process is possible from the view point of heat balance.

The temperature distribution in the vessel should be a main factor of the formation of a metal pool, because it was shown that the sufficient heat is supplied in overall.   It was reported that the highest temperature in the ordinary ESR operation was at the tip of consumable electrode.[11]   In order to get high temperature on the base-plate, strong fluid flow should be necessary in the melt.

The Peltier's heat for the reaction (1) should also be investigated. Further study is necessary on the temperature distribution.

Conclusion

By the addition of Al_2O_3 to the slag, the electro-deposit of Ti-Al was obtained. The current efficiency increased with the supplied electric power. It is concluded that the mechanism of high current efficiency should be the formation of liquid metal deposit due to high heat generation and the suppression of metal fog formation and sub-oxide formation of titanium.

References

1) George Zheng Chen, Derek J. Fray and Tom W. Ferthing: *NATURE*, 407, (2000) 361-344.
2) K. Ono and R. Suzuki: *J. Minerals, Metals & Materials Society (JOM)*, Feb., (2002) 59-61.
3) M. Kawakami, M. Ooishi, T. Takenaka and T. Suzuki: *Proc. 5th Intnl. Conf. Molten Slags, Fluxes and Slags '97*, (1997) 477.
4) M. Kawakami, and T. Kusamichi: *Japanese Patent, H06-146094*, (1994) (in Japanese).
5) T. Takenaka, T. Suzuki, M. Ishikawa, E. Fukasawa, and M. Kawakami: *Electrochemistry*, 67, (1999) 661-668.
6) M. Kawakami, T. Takenaka, M. Orisaka, T. Kawabata, A. Matsuyama and S. Yokoyama: *Proc. Sohn International Symposium ADVANCED PROCESSING OF METALS AND MATERIALS*, San Diego, TMS, vol.7, pp.373-381.
7) A. Mitchell. and J. Cameron: *Metall. Trans.*, 2, (1971) 3361-3366.
8) A. I. Zaitsev and B.M. Mogutnov, *Metallurgical and Materials Trans.*, 32B, (2001) 305-311.
9) I. Barin and O Knacke, *"Themochemical properties of Inorganic Substances"*, Springer Verlag, Berlin, (1973).
10) A. Mitchell and S. Joshi, *Metall. Trans.*, 4, (1973) 631.
11) M. Kawakami, K. Nagata, K. Yamamura, N. Sakata, Y. Miyashita, and K.Goto, *Tetsu-to-Hagane*, 63, (1977) 2162-2171.

Acknowledgement

The present work was supported by the Ministry of Education, Culture, Sports, Science and Technology, Japan (Grant-in-Aid for Scientific Research (A)(2), #15206084).

A NOVEL RECYCLING PROCESS OF TITANIUM METAL SCRAPS BY USING CHLORIDE WASTES

Haiyan Zheng[1] and Toru H. Okabe[2]

1 Graduate School of Engineering, The University of Tokyo;
7-3-1 Hongo, Bunkyo-ku, Tokyo 113-8656, Japan
2 Institute of Industrial Science, The University of Tokyo;
4-6-1 Komaba, Meguro-ku, Tokyo 153-8505, Japan

Keywords: Recycling, Titanium metal scraps, Chloride wastes

Abstract

A novel process of recycling titanium metal scraps by utilizing chloride wastes (e.g., $FeCl_x$ and $AlCl_3$) that are obtained as by-products in the Kroll process or any other chlorination process has been investigated in this study. This is important from the viewpoint of the increase in titanium metal scraps and chloride wastes in the future. Thermodynamic analyses and some primary experiments have been carried out in previous studies. Based on these studies, a fundamental study was carried out: titanium granules were reacted with iron chloride ($FeCl_2$) in a sealed quartz tube under a reduced argon atmosphere over a temperature range of 900–1300 K. The analytical results of the obtained samples determined using inductively coupled plasma-atomic emission spectrometry (ICP-AES), the potentiometric titration method, and X-ray diffraction (XRD) analyses reveal that the chlorine contained in the chloride wastes can be recovered by titanium granules as titanium chloride and the obtained titanium chloride can be easily separated from iron and other chlorides and then recovered. The recovered titanium chloride can be used as the feed material in titanium smelting processes.

Introduction

Currently, the titanium production process is referred to as the Kroll process [1–5], which basically involves the following three major chemical reaction steps: (1) carbo-chlorination of titanium oxide for producing titanium tetrachloride ($TiCl_4$), (2) purification of the obtained $TiCl_4$ by distillation, and (3) magnesiothermic reduction of the purified $TiCl_4$. Although this process can yield high-purity titanium, its production cost is high. This is partly because the reduction process is extremely slow and it involves an inefficient batch-type process. Furthermore, some amount of chloride wastes such as iron chlorides ($FeCl_x$, x = 2, 3) are generated from the chlorination process, as shown in Figure 1, because the titanium oxide feed contains impurities such as iron. At present, there is no process that efficiently recycles or treats the chloride wastes that are generated from the Kroll process; therefore, they are discarded after chemical treatment. The treatment of chloride wastes involves several problems such as the disposal cost and environmental issues, particularly in Japan. In addition, an additional amount of chlorine gas has to be purchased in order to compensate for the chlorine loss caused during the generation of chloride wastes. In order to minimize the generation of chloride wastes, rutile ore or upgraded ilmenite (UGI) comprising approximately 95% or more of titanium oxide is currently used as the raw material in the Kroll process. In the future, the amount of chloride wastes will increase as the production volume of titanium increases. It is projected that the quantity of titanium metal scraps will also increase in the future. These metal scraps are currently used for ferro-alloys, but it would be advantageous if they could be reused in the titanium smelting process for producing pure titanium.

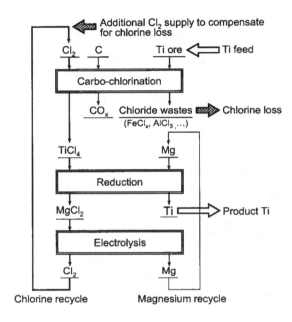

Figure 1 Chlorine cycle in the Kroll process.

Due to the abovementioned factors, the authors are currently investigating a new process, as shown in Figure 2. The chlorine and titanium recovery process investigated in this study is a part of the new process that is shown as "2: Chlorine recovery" in Figure 2. In the process investigated in this study, titanium metal scraps are recycled by utilizing the chloride wastes generated from the upgrading process of low-grade titanium ore or from the Kroll process. If this new process is feasible, not only can the titanium metal scraps be recycled but it would also be possible to effectively recover the chlorine in the chloride wastes generated from titanium smelting or any other process. Another benefit of this process is that a low-grade titanium ore can be used in the Kroll process if chlorine can be efficiently recovered. In addition, this recycling process in which a combination of titanium metal scraps and chloride wastes is used can also be extended to other reactive metals such as rare earth metals and tantalum [6].

148

In order to establish this new environmentally sound recycling process, the thermodynamic analyses of the reactions between $FeCl_x$ and Ti and its oxides (TiO_x) and some preliminary experimental works were carried out in the previous study [7]. In that study, titanium powder was used as the initial material and it was reacted with $FeCl_2$ at 1100 K for 1 h. After the experiment, the concentrations of Ti and Fe in the obtained sample were 9.9% and 90.1%, respectively. It is demonstrated that it is feasible to recycle titanium metal scraps by utilizing chloride wastes as a chlorine resource. On the basis of the results of the previous studies, the experiments using titanium granules (particle size: 1–2 mm) as the starting material are conducted in this study, because it is more practical to use coarse titanium scraps as a starting material.

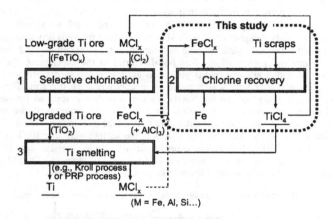

Figure 2 Flowchart of the new process discussed in this study.
1. Selective chlorination;
2. Chlorine and titanium scrap recovery;
3. Ti smelting, e.g. Kroll process and PRP process.

149

Experimental

The study for recycling titanium metal scraps with chloride wastes is based on the following reaction with an excess amount of $FeCl_2$ [7, 8].

$$Ti\ (s) + 2\ FeCl_2\ (l) \rightarrow TiCl_4\ (g) + 2\ Fe\ (s) \qquad (1)$$
$$\Delta G^\circ = -205\ kJ\ at\ 1100\ K$$

The experimental work for obtaining titanium chloride was carried out by reacting metallic titanium and $FeCl_2$. Figure 3 shows a schematic illustration of the experimental apparatus used for the chlorination of metallic titanium. In a typical experiment, a graphite crucible (I.D. ϕ = 27 mm) was filled with a mixture of titanium powder (31 μm) and $FeCl_2$ powder, following which the crucible was placed in a quartz tube (I.D. ϕ = 41 mm and length = 450 mm). A NaOH gas trap sustained in a glass flange covered with SUS nets on both the ends was also installed in the quartz tube near the gas outlet port. In some experiments, titanium granules (1–2 mm) were used as the starting material instead of titanium powder. Prior to the experiment, the quartz tube, which was sealed with a silicone rubber plug, was evacuated and then filled with Ar gas; the pressure within the quartz tube was maintained at approximately 0.2 atm. The quartz tube containing the sample mixture and the NaOH gas trap was then introduced into an isothermal horizontal furnace and maintained at 1100 K for 3 h. After the experiment, the residue in the graphite crucible and the deposits both within the quartz tube and on the surface of NaOH were recovered and subsequently analyzed. The phases in the sample were identified by using X-ray diffraction analysis (XRD). Inductively coupled plasma-atomic emission spectrometry (ICP-AES) was used for determining the concentration of the metallic elements. The chlorine element in the sample was determined by using the potentiometric titration method.

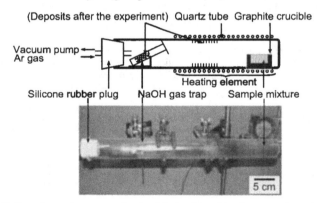

Figure 3 Experimental apparatus for the chlorination of titanium granules using $FeCl_2$.

Table I Initial materials used in this study.

Materials	Form	Purity (%)	Note / Supplier
Ti	Powder	98.0 up	Toho Titanium Co., Ltd.
Ti sponge	Granule	99.2*	Toho Titanium Co., Ltd.
$FeCl_2$	Powder	99.0	Kojyundo Chemical Laboratory Co., Ltd.

*: Determined by X-ray fluorescence analysis (XRF).

150

Results and discussion

Figure 4 (a) shows the assembled quartz tube after the experiment; the mixture of titanium granules and $FeCl_2$ was transferred to this tube and maintained at 1100 K for 3 h. During the experiments, the evolution of a white vapor, which was considered to be $TiCl_4$ or its related compounds, was observed. As shown in Figure 4, the major portion of this white vapor was deposited on the NaOH gas trap, whereas a small portion was deposited inside the quartz tube. After heating, brown flakes were deposited inside the quartz tube at a distance of 25 cm from the outlet of the quartz tube (Figure 4 (c)), and the residue (Figure 4 (d)) obtained in the graphite crucible was covered with a black powder.

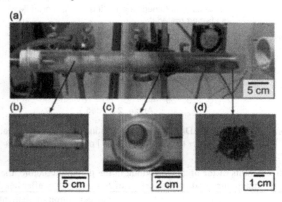

Figure 4 (a) Assembled quartz tube after experiment;
 (b) Deposit on the surface of the NaOH gas trap;
 (c) Deposit inside the quartz tube;
 (d) Residue in the graphite crucible.

Figure 5 shows the XRD patterns of the residues obtained after heating with titanium powder and titanium granules as the starting material. The XRD patterns indicate that α-Fe was obtained in the sample after heating. However, when titanium granules were used as the starting material, the amount of unreacted Ti increased after heating. This indicates that the efficiency of the recovery of titanium scraps and chlorine was effected by the morphology of the titanium scraps. As compared to the results obtained in the previous study [7], the speed of chlorine formation was slow, and the reaction speed decreased when titanium granules were used.

The results of the element analysis, as determined by ICP-AES and potentiometric titration, are listed in Table II. The black powder on the surface of the obtained sample was considered to be Fe powder according to the analytical results shown in Table II, the XRD analysis, magnetic properties, and color of the powder. As shown in Table II, the flakes that were deposited at a distance of 25 cm from the outlet of the quartz tube were $FeCl_2$, which was transferred from the graphite crucible as vapor and was condensed and deposited inside the quartz tube. The vapor pressure of $FeCl_2$ at the reaction temperature (1100 K) is approximately 0.093 atm, and the temperature at which deposition occurs (at a distance of 25 cm from the outlet of the quartz tube) is approximately 990 K and the vapor pressure of $FeCl_2$ at this temperature is approximately 0.02 atm.

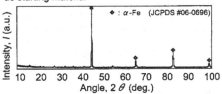

(a) Residue obtained by using titanium powder as starting material

(b) Residue obtained when using titanium granules as starting material

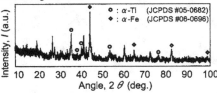

Figure 5 Comparison of the XRD patterns of the residue obtained after heating the sample mixed with either titanium powder [7] or titanium granules

Table II Analytical results of the obtained residue in the graphite crucible, the deposits obtained inside the quartz tube and on the surface of NaOH gas trap after heating.

Exp. # CD	Concentration of element i, C_i (mass%)		
	Ti	Fe	Cl
Residue in the graphite crucible	62.8[a]	37.2[a]	-[a]
Deposit inside the quartz tube	0.1[a]	49.7[a]	50.2[b]
Deposit on the surface of NaOH gas trap	0.2[a]	0.4[a]	99.4[b]

a: Determined by inductively coupled plasma-atomic emission spectrometry (ICP-AES).
b: Determined by the potentiometric titration method.

The abovementioned results demonstrated that the recovery of chlorine in $FeCl_2$ by using titanium granules is feasible. However, the efficiency of the recovery of titanium and chlorine from the mixture of titanium scraps and $FeCl_x$ is still low, and the efficiency needs to be improved for practical applications. Furthermore, the reaction speed was lower than that using titanium powder as the starting material [8]. Currently, the authors are investigating the mass balance of the chlorination reaction (Equation 2) and studying a new method for enhancing the efficiency of recovery and the reaction speed. Some recycling processes of other reactive metal scraps by using chloride wastes are also under investigation.

$$Ti\ (s) + m\ FeCl_x\ (l, g) \rightarrow TiCl_y\ (g) + n\ Fe\ (s) \qquad (2)$$

152

Conclusions and remarks

Some fundamental experiments on recycling chlorine present in chloride wastes by utilizing titanium metal scraps were carried out. It was demonstrated that the chlorine contained in the chloride wastes can be recovered by using titanium granules at 1100 K. The efficiency of titanium and chlorine recovery and the reaction speed were lower when titanium granules were used as the starting material as compared to the case in which titanium powder was used as the starting material. For enhancing the recovery efficiency and the reaction speed, the mass balance of the chlorination reaction of titanium metal scraps is under investigation.

Acknowledgments

The authors are grateful to Professors M. Maeda, Y. Mitsuda, and K. Morita of the University of Tokyo; Professor T. Uda of Kyoto University; and Messrs. S. Kosemura, M. Yamaguchi, and Y. Ono of the Toho Titanium Co., Ltd. for their valuable discussions and the samples supplied during the course of this study. Further, we thank Messrs. I. Maebashi and O. Takeda (currently at Tohoku University, Japan) for their valuable suggestions and technical assistance. We would like to specially thank Mr. R. Matsuoka of Cabot Supermetals K.K. for his preliminary study. A part of this research has been financially supported by a Grant-in-Aid for Young Scientists (A) from the Ministry of Education, Culture, Sports, Science and Technology (MEXT, Project ID. #16656232).

One of the authors, H. Zheng, is grateful to the Japan Society for the Promotion of Science (JSPS) Core-to-Core Program for the FY2005–2006 (Project ID. #17002 Development of Environmentally Sound Active Metal Process) for providing financial support to attend this conference.

References

1.   W. Kroll, "The Production of Ductile Titanium," *Tr. Electrochem. Soc.*, 78 (1940), 35–47.

2.   T. Fukuyama et al., "Production of Titanium Sponge and Ingot at Toho Titanium Co. Ltd.," *Journal of the Mining and Materials Processing Institute of Japan,* 109 (1993), 1157–1163.

3.   A. Moriya and A. Kanai, "Production of Titanium at Sumitomo Sitix Corporation," *Journal of the Mining and Materials Processing Institute of Japan*, 109 (1993), 1164–1169.

4.   F. Habashi, *Handbook of Extractive Metallurgy*, (Weinheim, Federal Republic of Germany: VCH Verlagsgesellschaft mbH, 1997), 1129–1180.

5.   The Japan Titanium Society ed., "Titanium is now taking off from the cradle to the growing age" (Commemorative Publication of 50th Anniversary of JTS), (Tokyo, Japan: The Japan Titanium Society, 2002), 47.

6.   K. Mineta and T.H. Okabe, "Development of Recycling Process for Tantalum from Capacitor Scraps," *Journal of Physics and Chemistry of Solids,* 66 (2005), 318–321.

7.   H. Zheng, R. Matsuoka, and T.H. Okabe, "Recycling Titanium Metal Scraps by Utilizing Chloride Wastes," *Proc. European Metallurgical Conference 2005, EMC2005* [Dresden, Germany, September 18–21, 2005], 1509–1518.

8.   I. Barin, *Thermochemical Data of Pure Substances, 3rd ed.,* (Weinheim, Federal Republic of Germany: VCH Verlagsgesellschaft mbH, 1997).

Innovations in
Titanium
Technology

Advances in Materials Processing

ADVANCES IN TITANIUM METAL INJECTION MOLDING

F. H. (Sam) Froes

Institute for Materials and Advanced Processes (IMAP),
University of Idaho, Moscow, ID 83844-3026

Keywords: Titanium, Metal Injection Molding, Binders, Oxygen Content, Mechanical Properties

Abstract

The approaches to production of titanium powder injection molded parts are reviewed. Although cosmetic (i.e. non-load bearing) parts have been successfully fabricated from commercially pure titanium historically, oxygen levels have been too high (because of degraded ductility) for structural use, particularly with the Ti-6Al-4V alloy. However, recent advances in starting powders, binders and sintering facilities now allow oxygen levels in the Ti-6Al-4V alloy to be controlled to about 0.2 wt. % oxygen. This should result in significant expansion of the titanium MIM market place into aerospace, automobiles, surgical instruments, dentistry, communication devices (such as computers and cell phones), knives and guns.

Introduction

Advanced materials are key to enhanced behavior in aerospace and terrestrial applications [1-2], and titanium alloys are amongst the most important of the advanced materials because of their excellent combinations of specific mechanical properties (properties normalized by density) and outstanding corrosion behavior [3-8]. However, a major concern with titanium alloys is their high cost compared to competing materials (Table I), which has led to investigation of various near net shapes (NNS) potentially lower cost processes, including powder metallurgy (P/M) and casting techniques [3-14].

In this paper, a brief overview of the metallurgy of the titanium system will be presented followed by a status report of the NNS approach of metal injection molding (MIM) to shape making. MIM is an approach of choice when a large number of small parts (less than 1 lb) of highly complex configuration are required, (Figs. 1 and 2).

*Table 1: Price of Titanium – A Comparison**

ITEM	MATERIAL ($ PER POUND)		
	STEEL	ALUMINUM	TITANIUM
ORE	0.02	0.10	0.22 (RUTILE)
METAL	0.10	0.10	5.44
INGOT	0.15	1.15	9.07
SHEET	0.30 – 0.60	1.00 – 5.00	15.00 – 50.00

*Contract prices. The high cost of titanium compared to aluminum and steel is a result of (a) high extraction costs and (b) high processing costs. The latter relating to the relatively low processing temperatures used for titanium and the conditioning (surface regions contaminated at the processing temperatures, and surface cracks, both of which must be removed) required prior to further fabrication.

157

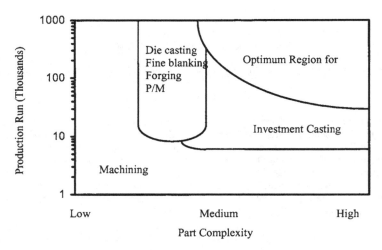

Figure 1: Diagram showing where Ti MIM is most appropriately used in comparison with other fabrication processes (Courtesy of Krebsöge Radevormwald).

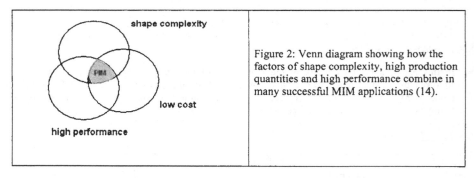

Figure 2: Venn diagram showing how the factors of shape complexity, high production quantities and high performance combine in many successful MIM applications (14).

The mechanical and physical properties of titanium systems depend on chemistry (Fig. 3) and microstructure (Fig. 4) (5-7). The alpha alloys are characterized by relatively low strength (≤ 80 ksi UTS), with a number of this class of alloys being used for high temperature applications (≤ 600°C).

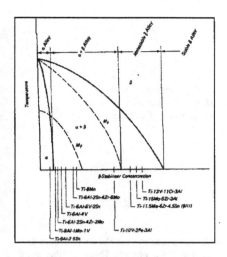

Figure 3: Compositions of U. S technical Alloys mapped onto a pseudobinary β-isomorphous phase diagram. (Courtesy ASM Int.)

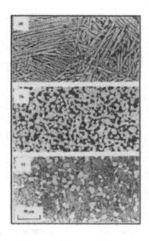

Figure 4: Microstructure of Ti-6Al-2Sn-4Zr-2Mo: (a) β worked followed by α-β anneal to produce lenticular α morphology, (b) α-β worked and α-β annealed to give predominantly an equiaxed α shape and (c) α-β worked followed by duplex anneal: just below the β transus temperature [reduced volume fraction of equiaxed α compared to (b)], and significantly below the β transus temperature (to form the lenticulary α between equiaxed regions) [7,8].

The alpha-beta class of alloys have higher strength in combination with reasonable levels of ductility, for example, the Ti-6Al-4V alloy exhibits minimums of 130 ksi UTS and 12% elongation. The beta alloys have strength equivalent to Ti-6Al-4V with significantly higher ductility. A class of titanium alloys not shown on Fig. 4 are the intermetallic Ti$_x$Al (x = 1 or 3) which have excellent high temperature behavior, but very low room temperature ductility (often 2% elongation maximum). Generally, titanium alloys increase in strength and decrease in ductility as the oxygen level is increased (max. aerospace specification for O_2 Ti-6Al-4V is 0.20 w/o), (Fig. 5).

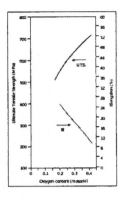

Figure 5: Effect of oxygen content on the strength and ductility of sintered commercially pure titanium powder (Courtesy Daido Steel Co.).

The two basic microstructures exhibited by conventional titanium alloys are shown in Figure 4, as equiaxed alpha and elongated alpha (in both cases the white phase). These exhibit good ductility and fatigue crack initiation on one hand and good fracture toughness and creep performance on the other hand. The middle photomicrograph is a structure designed to give a good combination of properties.

Titanium Applications

Applications for titanium alloys can be separated into two categories high strength structural (with good fracture toughness and fatigue behavior) and lower strength corrosion resistant (5-7). Generally the alpha-beta and beta alloys are used for the former application and the commercially pure grades for the latter use. Examples of titanium components which could be made using the MIM approach are shown in Figures 6-9.

Figure 6: Toyota automotive components produced using a BE P/M approach (Courtesy Toyota Central Research Laboratory).

Figure 7: Ti-6Al-4V fasteners produced using the BE P/M technique suitable for automotive use (Courtesy ADMA Products).

160

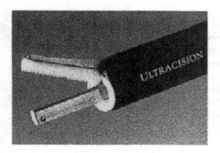

Figure 8: The ultracision harmonic scalpel which allows surgeons to make incisions with minimal tissue damage (Courtesy Johnson & Johnson).

Figure 9: Titanium alloy watchcase using the powder injection molding process (Courtesy Hitachi Metals Precision/Casio Computer Co.).

Titanium Powder Injection Molding

As shown in Figures 1 and 2, the use of MIM is favored by a large number of small complex parts. The sub-division can be made in the case of Ti MIM: cosmetic parts (where the mechanical properties are not important) and structural parts (exposed to stress, making mechanical properties of importance). A major contributor to the mechanical properties is the interstitial levels, particularly oxygen. Thus, the aerospace oxygen specification for CP titanium is 0.4 w/o, whereas for Ti-6Al-4V it is 0.2 w/o; with the former composition being used at lower strength (Grade 4) ($\leq$ 80 ksi) and the latter at levels of 130-140 ksi. Thus "cosmetic" parts such as watch cases are fabricated from CP titanium (Fig. 9).

The metal powder injection molding process is based upon the injection molding of plastics, a process developed for long production runs of small (normally below 400 gm) complex shaped parts in a cost-effective manner. By increasing the metal (or ceramic) particle content, the process evolved into a process for production of high density metal, intermetallic or ceramic components (Fig. 10) (12, 13).

161

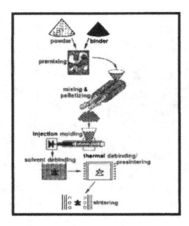

Figure 10: Schematic of the steps involved in powder injection molding, in which a polymer binder and metal powder are mixed to form the feedstock which is molded, debound and sintered. The process relies on the thermoplastic binder for shaping at a moderately elevated temperature of about 150°C [12,13].

Much of the early work on developing a viable titanium MIM process was plagued by the unavailability of suitable powder, inadequate protection of the titanium during elevated temperature processing and less than optimum binders for a material as reactive as titanium (13). However, some MIM practitioners have now learned what the titanium community has long known – that titanium is the universal solvent and must be treated accordingly [4-8].

Suitable powders are now available (Figs. 11-13) and sintering furnaces suitable for use with titanium are now in place. Thus, the challenge now is to find suitable binders. Unfortunately, even some of the more well known polymer binders known for their ability to readily thermally unzip to their starting monomers (e.g. polymethyl methacrylate, polypropylene carbonate, poly-α-methyl styrene) still tend to introduce impurities into the sintered Ti MIM bodies because their depolymerization occurs close to those temperatures where impurity uptake initiates (at ≈ 260°C). Alternative binder systems based on catalytic decomposition of polyacetals are promising, but require expensive capital equipment to handle the acid vapor catalyst as well as suitable means of eliminating the formaldehyde oligomers that form as polymer decomposition by-products. However, there are a number of binder systems which appear to have the necessary characteristics to be compatible with titanium (Table II) giving acceptable levels of oxygen content, strength and ductility (Table III). Factors which affect the strength – ductility are the oxygen level (strength up, ductility down), relative density (strength and ductility up) and the beta grain size (smaller grain size for increased strength and ductility). An additional effect is powder size, with a smaller size likely to give increased oxygen content and a finer beta grain size.

162

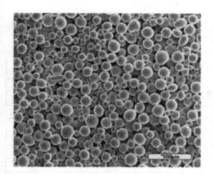

Figure 11: Gas atomized prealloyed spherical Ti-6Al-4V (Courtesy of Affinity International).

Figure 12: SEM photomicrograph of the powder produced from Ti sponge fines hydrogenated at 1400°F.

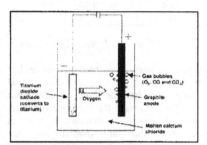

Figure 13a: Schematic of the Fray electrolytic process for producing titanium.

Figure 13B: An SEM of the Fray titanium product (Courtesy G. Chen, Cambridge University).

*Table II. Binder systems which appear to be compatible *with Ti-6Al-4V*

Polypropylene-Ethel Vinyl Acetate-Paraffin Wax-Carnauba Wax-Dioctyl Phthalate (15))
Polyethylene, Paraffin And Stearic Acid (16)
Polypropylene-Polymethy Methacrylate-Paraffin-Stearic Acid (17)
Polypropylene-Paraffin-Carnauba, Etc. (18)
Secret (19)
Naphthalene-Stearic Acid-Ethylene Vinyl Acetate (20)
Paraffin Wax – Polyethylglycol – Polyethylene – Stearic Acid (21)
Paraffin Wax – Co-Polymer – Stearic Acid (22)
Atactic Polypropolene – Carnauba Wax – Paraffin Wax – Stearic Acid (23)
PP – EVA – PW – CW – DOP (24)
Specially Developed on Polymer Base (26)
Atactic polypropylene-ethylene vinyl acetate-paraffin-carnauba-Di-n-butyl phthalate (27)

*See Table III

Table III. Characteristics of Ti MIM*

Oxygen content (w/o)	Relative Density (%)	UTS (ksi)	Elongation (%)	Reference
0.24	96.0	140.8	12	15
0.28	97.1	118.3	7.8[1]	16
0.25 – 0.28	95.5	121.9	14.0	17
0.20	95.1	94.3	22[2]	18
0.19	~95%	–	–	19
0.17	99.5[3]	152.1	14.6	20
0.54	96.7	121.9	9.0	21
–	99.5	136.1	14.0	22
0.32	>96.0	136.4	2.5	23
0.34	96	139.3	11.2	24
0.24	98	133.5	14.0[4]	25 (17)
0.25	95.5	121.2	13.4	26
0.26	96.0	128	10	27

* Ti-6Al-4V unless stated
1. Ti-6Al-7Nb   2. Commercially pure Titanium   3. HIP'd   3. Near Alpha alloy

Currently titanium MIM parts run up to a foot in length, but parts over three or four inches (about 50 gm in weight) are not common. The limiting factors at this time are dimensional reproducibility and chemistry. Due to the shrinkage, large parts become dimensionally more difficult to make due to loss of shape during shrinkage. If the parts have flat surfaces to rest on the setter they come out fairly consistently. But parts with multiple surfaces that require setters in complex shapes become less practical as the size goes up. Further, large overhanging areas become difficult to control dimensionally due to gravity. With increasing experience, the packing density of titanium powder mixes will be increased, especially with the new binders that are now becoming available and the shrinkage can be decreased making the dimensional problems less of a factor.

The Future

The current estimate is that the world-wide titanium MIM part production is currently at about the 3 to 5 ton per month level. This market is poised for expansion. What is needed is low cost (less than $20/lb or $44/kg) powder of the right size (less than about 40 microns) and good purity (which is maintained throughout the fabrication process). For non-aerospace applications, the purity level of the Ti-6Al-4V alloy can be less stringent; for example, the oxygen level can be up to 0.3 wt% while still exhibiting acceptable ductility levels (aerospace requires a maximum oxygen level of 0.2 wt% [8]). For the CP grades, oxygen levels can be even higher; up to at least 0.4 wt. % (Grade 4 CP titanium has a spec. limit of 0.4 wt% [8]). In fact, the Grade 4 CP titanium (UTS 550 MPa [80 ksi]) while lower strength than regular Ti-6Al-4V (UTS 930 MPa [135 ksi]) may well be a better choice for many potential PIM parts where cost is of great concern. The Grade 4 would allow use of a lower cost starting stock and a higher oxygen content in the final part. Further into the future, the beta alloys with their inherent good ductility (bcc structure) and the intermetallics with attractive elevated temperature capability are potential candidates for fabrication via MIM. The science, technology and cost now seem to be in place for the titanium MIM marketplace to show significant growth.

A variety of high quality, relatively low cost powders are now available. There have also been a number of developments, including development of suitable binders and sintering furnaces, which should lead to a reasonable growth of titanium products produced by the MIM method. The big growth potential is in small complex shaped parts using the MIM approach (Figs. 1 and 2). Early entrants to this market place naively largely ignored what every good titanium metallurgist knows – that titanium is the universal solvent. Hopefully, this fact is now clear to current and pending titanium MIM practitioners. With the production of high integrity (particularly oxygen within specification limits) cost effective, complex MIM components a market in both aerospace and terrestrial industries should grow quite rapidly. A particular target should be automobile use – there are approximately 15 million cars and light trucks built in the USA alone each year.

Conclusions

The approaches which have been taken to produce titanium MIM parts have been reviewed.
Parts produced have historically been high in oxygen preventing their use in structural (load bearing) applications; however, cosmetic parts (not exposed to any stress) have been successfully produced mainly from commercially pure grades of titanium (which allow up to 0.4 wt %).
It is concluded that starting powders of suitable quality and price, along with sintering furnaces which minimize oxygen pick up, and a number of binders which do not result in significant oxygen pick up are now available.

Thus, there is a high potential for production of structural titanium MIM parts in industries such as aerospace, automobiles, surgical instruments, dentistry, communication devices (computers, cell phones, etc.), knives and guns.

Acknowledgements

The authors recognize useful discussions with Serge Grenier, Joe Grohowski, Andy Hanson, John (Qiang) Li, Tim McCabe, Eric Nyberg, Kevin Simmons and Fred Yolton. We also acknowledge the contribution of Ms. Linda Shepard in formatting and typing the text.

References

1.   Congress of the U.S. Office of Technology Assessment, Advanced Materials by Design, June 1988.

2.   Materials Science and Engineering – Forging Stronger Links to Users, 1999, NMAB, National Academy Press publication NMAB-492, Washington DC.

3.   F.H. Froes, D. Eylon, and H. Bomberger eds., "Titanium Technology: Present Status and Future Trends", *TDA*, Dayton, OH, 1985.

4.   Francis H. (Sam) Froes, Te-Lin Yau and Hans G. Weidenger, "Titanium, Zirconium and Hafnium" Chapter 8, *Materials Science and Technology – Structure and Properties of Nonferrous Alloys*, vol. ed. K.H. Matucha, 1996, VCH Weinheim, FRG, p. 401.

5.   F.H. (Sam) Froes, "Titanium", Chapters 3.3.5a – 3.3.5e *Encyclopedia of Materials Science and Engineering*, P. Bridenbaugh, subject editor, Elsevier, Oxford, UK, 2001 pp. 9361-9374.

6.   F.H. (Sam) Froes, "Titanium Alloys", chapter 8 of the *Handbook of Advanced Materials*, Ed. in Chief, James K. Wessel, Wiley Interscience, (2004), p. 271.

7.   F.H. (Sam) Froes "Titanium Metal Alloys", *Handbook of Chemical Industry Economics, Inorganic*, Ed. in Chief Jeff Ellis, John Wiley and Sons Inc., New York, NY. To be published 2002.

8.   R.R. Boyer, G. Welsch and E.W. Collings, Eds. "Materials Properties Handbook: Titanium Alloys", 1994, *ASM Int.*, Materials Park, OH.

9. F.H. Froes and D. Eylon, "Powder Metallurgy of Titanium Alloys," *International Materials Reviews*, Vol. 35, 1990, p. 162.

10. F.H. Froes and C. Suryanarayana, "Powder Processing of Titanium Alloys," Reviews in Particulate Materials, Eds. A. Bose, R.M. German and A. Lawley, MPIF, Princeton, NJ, vol. 1, 1993, p. 223.

11. F. Arcella and F.H. (Sam) Froes, "Production Of Titanium Aerospace Components From Powder Using Laser Forming", JOM Vol. 52, No. 5 (2000), p. 28.

12. Powder Metallurgy Science, 2nd Edition, R.M. German, 1994, Chapter Six, p. 192 et seq. MPIF, Princeton, NJ.

13. F.H. (Sam) Froes and R.M. German, Metal Powder Report, "Titanium Powder Injection Molding (PIM)", Vol. 55, No. 6 (2000), p. 12.

14. Randall M. German, "Powder Injection Molding Design and Applications – Users Guide" Innovative Solutions, State College, PA 16803, USA, 2003, p. 5

15. Tonio Kono, Akira Horata and Tetsuya Kondo, "Development of Titanium & Titanium Alloy by Metal Injection Molding Process", Powder and Powder Metallurgy, Vol. 44, No. 11 (in Japanese).

16. W. Limberg, E. Aust, T. Ebel, R. Gerling and B. Oger, "Metal Injection Molding of an Advanced Bone Screw Ti-6Al-7Nb Alloy Powder", Euro 2004.

17. Yoshinri Itoh, Tatsuya Hankou, Kenji Sato and Hideshi Miura, "Improvement of Ductility for Injection Moulding Ti-6Al-4V Alloy", Euro 2004.

18. Hidefumi Nakamura, Tokihiro Shimura and Kouei Nakabayashi, "Process for Production of Ti Sintered Compacts Using the Injection Molding Method", J. Jap. Soc. Of Powder and P/M, Vol. 46, No. 8 (1999), (in Japanese).

19. Private communication with J. Grohowski, Praxis Technology, Feb. 27, 2006.

20. Kevin Simmons, K. Scott Weil and Eric Nyberg, "Powder Injection Molding of Titanium Compounds", Industrial Heating, Dec. 2005, p. 43.

21. Guo Shibo et al. "Influence of Sintering Time on Mechanical Properties of Ti-6Al-4V Compacts by Metal Injection Molding", Rare Metal Materials and Engineering, Vol. 34, No.7 (July 2005), p. 33.

22. Hermina Wang et al, "Development of High density (99% +) Powder Injection Molded Titanium Alloys" PIM Science and Technical Briefs, Vol. 1, No. 5 (1999), p. 16.

23. Keiiclu Maekawa et al. "Effect of MIM Process Conditions on Microstructures and Mechanical Properties of Ti-6Al-4V Compacts", J. Japan Soc. of P/M Vol. 46, No. 10 (1999), p. 1053.

24. Katsushi Kusaka et al. "Tensile Behavior of Sintered Ti and Ti-6Al-4V Alloy by MIM Process", Advances in PIM and Particulate Materials, MPIF, Princeton NJ (1996), p. 29-127.

25. Yoshinori Itoh et. al. "Fabrication of Near Alpha Titanium Alloy by Metal Injection Molding", J. Japan P/M, Vol. 52, No. 1, (2005), p. 43.

26. Gerhard Wegmann et al. "Metal Injection Molding of Titanium Alloys for Medical Applications", Materials Week 2000, p. 1.

27. Makolo Fujita et al "New Process for Ti Alloy Powders Production Using Gas Atomization" Korean World P/M Conf. Sept. 2006.

Innovations in **Titanium Technology**

Advances in Alloy Development

PROCESSING AND PROPERTY IMPROVEMENTS IN ROLLED PLATES AND SHEETS OF TI-6AL-4V + 0.1 WT% B

Mats Bennett[1], Raghavan Srinivasan[1], Sesh Tamirisa[2]

[1] Department of Mechanical and Materials Engineering, Wright State University
3640 Colonel Glenn Hwy.; Dayton, OH 45435, USA
[2] Department of Mechanical Engineering, Ohio University, Athens, OH 45701, USA

Keywords: TiB, Ti-6Al-4V, Rolling

Abstract

The effect of boron addition on the deformation behavior and mechanical properties of a Ti-6Al-4V alloy was studied. Samples of 25 mm thickness were sectioned out of ingots of Ti-6Al-4V+0.1wt%B ingots prepared by ISM (Induction Skull Melting) and PAM (Plasma Arc Melting) and rolled at 954°C (1750°F) and 982°C (1800°F) to 6 mm plate (75% reduction in thickness) and 2 mm sheet (92% reduction in thickness). Microstructures were obtained from the as-cast material as well as the rolled material. In the as-cast condition, the prior beta grain size was in the range of 200-250 μm. TiB whiskers were found to decorate the prior beta grain boundaries. Microstructures of the rolled samples showed the TiB whiskers breakup, and the matrix fills the resulting gaps between TiB particles. Room temperature tensile tests conducted on both plate and sheet samples showed increases in yield strength, ultimate tensile strength and ductility compared to conventionally prepared Ti-6Al-4V. Heat treatments were conducted on the 2 mm thick sheet samples for recrystallization and globularization. Metallographic examination showed a globular microstructure with a grain size in the range of 3 to 5 μm, making this material potentially useful for superplastic forming (SPF) applications.

Introduction

Titanium is the fourth most abundant metal in the earth's crust. The density of titanium is 4.5 g/cm³. Compared with metals such as iron (7.86 g/cm³) and nickel (8.9 g/cm³) this makes titanium very attractive in applications requiring low weight and high strength. Titanium is one of the most corrosion resistant metals in use today. The cost of titanium limits its applications in general consumer production [1,2,3,5,6].

The widest use of titanium is in the aerospace industry. Historically the aerospace industry and military have accounted for over 75% of titanium usage. The aerospace industry has been employing titanium for many years now, taking advantage of its low density and high strength. Titanium alloys such as Ti-6Al-4V (referred to as Ti-64 hereafter) display good diffusion bonding characteristics which aids in its ability to be superplastically formed (SPF). By this method, complex structures are then able to be created [1,2,5,6].

While the aerospace and military applications are predominantly performance driven, consumer industries have also begun to take advantage of the properties of titanium. Increased attention from the automotive industry, chemical industry, as well as the medical industry has drawn more interesting into the development of titanium alloys for commercial use. With the advent of consumer industry involvement the focus now is also on the development of cost effective alloys [2].

The most widely used titanium alloy in the market is Ti-64 (Ti-6Al-4V). Ti-64 accounts for 50% of the use of titanium in the world today. Ti-64 is the most widely used titanium alloy due to its high strength at temperatures up to 400 °C, corrosion resistance and relative ease of formability. Titanium sponge is initially extracted from the ore and is then purified. The sponge is then converted into briquettes which are welded together to form an electrode for melting. Typically vacuum arc remelting (VAR) is carried out to melt the electrode down and produce the cast Ti-64 ingot. After casting, Ti-64 has a grain size of several millimeters to centimeters depending on ingot size. This large grain size is detrimental to the properties of Ti-64. Ti-64 is consequently subjected to an ingot breakdown procedure to reduce overall grain size as well as create a workable product for secondary forming operations such as rolling or forging [1,2,4].

Through the work of the Air Force Research Labs (AFRL), boron was found to be a potent grain refiner for Ti-64. Figure 2 shows the effect of boron on grain size based on the weight percent added. The addition of 0.1 weight percent boron reduces the as-cast grain size of Ti-64 to 250 μm which is comparable to the size of the Ti-64 grains after ingot breakdown [4].

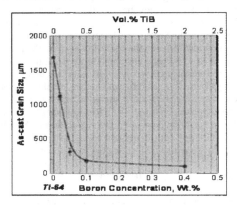

Figure 1: Effect of boron on grain size in Ti-64.

Reduction in grain size allows the elimination of steps such as ingot breakdown reducing the production cost of titanium. Specifically, the effects of rolling the ingot into plate and sheet will be studied showing subsequent microstructure as well as mechanical properties after deformation.

Experimental Procedure

Ti-64 as well as Ti-64- 0.1 wt%B cast by different procedures was obtained from RMI (Niles, OH) and Flowserve (Dayton, OH) as ingots. Three samples were obtained from PAM and one sample was obtained from ISM. Table 1 shows a summary of the chemical compositions of the different ingots.

Table 1: Summary of chemical compositions.

Piece ID	Melt Type	Composition	Piece Type
ISM 1	Induction Skull Melting	Ti-6Al-4V-0.1B	Transverse
5006	PAM Melt	Ti-6Al-4V-0.08O	Longitudinal
5010	PAM Melt	Ti-6Al-4V-0.1B-0.34O	Longitudinal
5011	PAM Melt	Ti-6Al-4V-0.1B-0.18O	Longitudinal

A micrograph from the as-cast condition for the PAM 5011 (Ti-64-0.1B) ingot is shown in Figure 2. In the PAM 5011 and ISM samples (Ti-64-0.1B) the grain size is on the order of 200 – 300 μm whereas in the conventional Ti-64 the overall grain size on the order of a few millimeters.

Figure 2: Ti-64-0.1B (PAM 5011) in the as-cast condition.

Rolling experiments were conducted on a laboratory rolling mill at RTI (Niles, OH). Samples measuring 127 mm (5 in) x 127 mm (5 in) x 25 mm (1 in) were cut from the existing PAM ingots (5006, 5010 and 5011) in the longitudinal direction. From the ISM sample an 89 mm (3.5 in) x 89 mm (3.5 in) x 25 mm (1 in) transverse sample was cut using EDM (electro-discharge machining). The gauge length of each sample was 19 mm with a thickness of 2.5 mm. The samples were rolled to 75% reduction in thickness using a laboratory rolling setup at RTI and their standard rolling procedure used in the company. A section of each rolled plate approximately 100 mm (4 inches) square was further rolled to 2 mm (0.085 inches) thickness for a total reduction of approximately 92% from the starting cast condition. The rolling of the plate was done at 954 °C (1750 °F). The rolling of the sheet was done at 982 °C (1800 °C). Each rolling pass was a reduction in thickness of approximately 10%.

Three tensile samples from the longitudinal direction as well as three tensile samples from the transverse direction were sectioned from the plate and sheet using EDM. Samples were surface polished to remove the EDM recast layer. Tensile testing was carried out at AFRL on a servo-hydraulic testing machine (MTS 810). An extensometer (MTS 632.31F-24) with 10 mm gauge length was attached to the specimen to obtain elongation. An automatic data collection software package (MTS Testworks4) was used to gather data as time, load and actuator displacement.

171

The data was then converted to engineering stress-strain curves using spreadsheet software (Excel). The nominal test speed was 0.01 mm/s and all tests were done at room temperature.

Results and Discussion

Microstructure

In the as-cast condition of a conventional Ti-64 ingot the grain structure is typically a lamellar alpha-beta microstructure with grain boundary alpha (α_{gb}) along the prior beta grain boundaries. For the Ti-64-0.1B ingot it can be seen from Figure 2 that TiB particles have replaced much of the α_{gb} and that the microstructure is also lamellar alpha-beta.

The microstructures for the 6 mm (0.25 in) rolled plate (75% reduction in thickness) can be seen in Figures 3 through #. It can be seen that the lamellar alpha-beta and TiB needles along prior-beta grain boundaries broke up during rolling. The grain size of the sample has been reduced to approximately 20 µm from 200-300 µm. The alpha now to shows a more globularized structure rather than the as-cast lamellar structure.

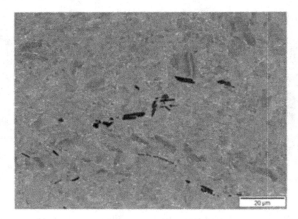

Figure 3: Micrograph of rolled plate of Ti-64-0.1B (PAM 5011) 75% reduction.

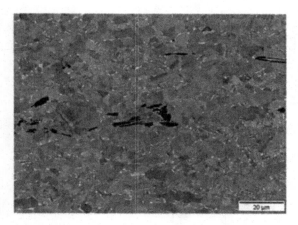

Figure 4: Micrograph of rolled plate of Ti-64-0.1B (ISM) 75% reduction.

The extent of globularization seems to be equivalent between the rolled plate and the heat treated sheet material for the 5011 sample. The overall grain size has been reduced greatly (20 μm to 5 μm) by the added deformation and subsequent heat treatment. Figures 5 and 6 show the microstructures of the rolled sheet (2 mm thickness).

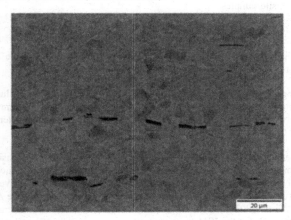

Figure 5: Micrograph of rolled sheet of Ti-64-0.1B (PAM 5011) 92% reduction.

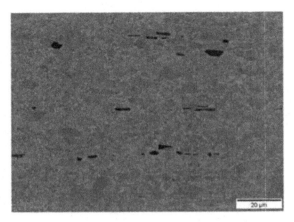

Figure 6: Micrograph of rolled sheet of Ti-64-0.1B (ISM) 92% reduction.

Another noticeable feature in the microstructures of the PAM and ISM plate and sheet is the distribution of TiB throughout the matrix. In the as-cast condition the TiB tended to group themselves along the prior beta grain boundaries in place of the normal grain boundary alpha. The cross-rolled TiB needles are broken up and distributed more and more evenly throughout the matrix with each successive pass. This may have had a significant effect upon the mechanical properties of the tensile samples.

Tensile Testing

A summary of the averages for the tensile test results is presented in Table 2. The average ductilities were obtained from measurements using a 19 mm gauge length.

Calculations using a 19 mm gauge length were used to compensate for failure of specimens outside of the extensometer. Incorporating the entire gauge length of the sample yielded more consistent ductility numbers. When failure occurred outside of the extensometer the curves have a distinct hook at the point of failure on the graph.

The ISM transverse samples had the highest yield stress and ultimate tensile stress. This observation is consistent with reported values for conventional Ti-64 in which the transverse direction has higher strength properties than the longitudinal direction. The average yield stress for the ISM transverse samples was 1021 MPa which when compared with the AMS 4911 value for conventional Ti-64 (903 MPa) is 13% better in performance. The average value of the ultimate tensile strength for the ISM transverse samples was 1060 MPa and when compared to standard Ti-64 (951 MPa) the boron modified samples outperformed the standard Ti-64 by 11%. These improvements in strength were not gained at the expense of ductility. The average ductility (average ductility was taken from the 19 mm gauge length calculations) for the ISM transverse samples was 20.5%.

The PAM 5011 transverse samples were the best performers as far as ductility was concerned. The average ductility was calculated to be 23% which is over twice the ductility required of standard Ti-64 per AMS specification. The strength values for the PAM 5011 transverse samples were close to conventional Ti-64 with an average yield strength of approximately 940

174

MPa. Figure 7 shows an example of the engineering stress versus engineering strain curve for the ISM plate.

Table 2: Summary of tensile test results.

Material	Direction	YS, MPa	UTS, MPa	Elon., %
Ti-6Al-4V	L	861	930	10
AMS 4911	T	903	951	10
Plate	L	945	987	18
PAM Ti-64-0.1B	T	934	983	23
Plate	L	986	1031	21
ISM Ti-64-0.1B	T	1021	1060	20
Sheet	L	953	1023	20
PAM Ti-64-0.1B	T	1038	1094	24
Plate	L	950	1033	25
ISM Ti-64-0.1B	T	1035	1081	23

For the heat treated sheet, the overall ductility improved for each specimen in both the final rolling direction as well as the transverse direction. The transverse direction had the best overall properties (Table 2). The PAM 5011 heat treated sheet tested in the transverse direction produced an average yield stress of 1038 MPa, an average ultimate tensile stress of 1094 MPa and an average ductility of 24.3%. The strengths for the 5011 heat treated sheet were significantly higher for the transverse direction when compared with that of the plate material. This could be a result of the cross rolling of the material. The last rolling direction does not necessarily represent the direction which received the most passes through the rolling mill.

For the ISM heat treated sheet tested in the transverse direction the numbers were similar to those of the non-heat treated 6 mm plate samples in the transverse direction. An average yield stress of 1035 MPa, an average ultimate tensile strength of 1081 MPa and an average ductility of 23.1% were recorded. Although the properties were not significantly better when compared to the 6 mm plate, the ductility was more consistent. This consistency was likely due to the heat treatment.

Like the plate, the average ductility across all the specimens was acceptable at 23%. Globularization as well as the heat treatment yielded higher values for yield strength, ultimate tensile strength and ductility. There is room for optimization during processing to improve the tensile properties.

175

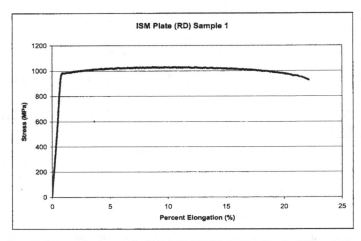

Figure 7: Stress vs. Strain curve for Ti-64-0.1B (ISM Final rolling Direction) 75% reduction.

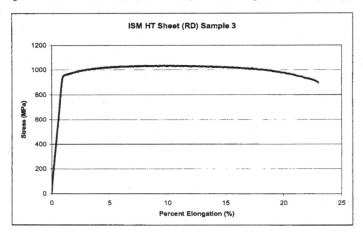

Figure 8: Stress vs. Strain curve for Ti-64-0.1B (ISM Final rolling Direction) 92% reduction.

Conclusion

The addition of 0.1% boron to Ti-64 reduces as-cast grain size by a factor of 10. This reduction in grain size allows for the rolling of Ti-64-0.1B directly from the ingot. Both the PAM and ISM ingots showed drastic improvements over conventional Ti-64. Cracks developed immediately following rolling to plate (6 mm) for the Ti-64 (Ingot 5006) sample. Conventional Ti-64 showed a lamellar alpha-beta microstructure with grain boundary alpha. The micrographs show a lamellar alpha-beta microstructure with TiB particles decorating the grain boundary for the Ti-64-0.1B samples.

Both ISM and PAM rolled plates equaled or outperformed conventional Ti-64 during tensile testing. The maximum yield stress displayed by the PAM 5011 plate samples was 955 MPa in the rolling direction with an ultimate tensile strength of 998 MPa and an elongation to failure of 14%. The ISM plate samples showed a maximum yield stress of 1023 MPa, a maximum ultimate tensile strength of 1063 MPa and an elongation to failure of 18% in the transverse direction. The formation of columnar grains during PAM casting resulted in a microstructure yielding slightly lower mechanical properties values for the transverse direction. Conversely, the ISM sample, which has an equiaxed microstructure in the as-cast condition, displayed better properties in the transverse direction. Compared with conventional Ti-64, the mechanical properties of Ti-64-0.1B are on average equivalent or slightly improved. Actual statistical significance could not be determined due to the low sample population during this experiment when compared with the AMS 4911 standard.

The reduction in grain size following the rolling to 92% reduction in thickness as well as the globularization following heat treatment resulted in higher elongations to failure as well as better yield strengths for the PAM 5011 sample. The PAM 5011 sheet produced a yield strength of 1043 MPa, an ultimate tensile strength of 1095 MPa and an elongation to failure of 24.3% in the transverse direction. Tests in the rolling direction showed yield strengths and ultimate tensile strength values approximately 100 MPa below the transverse strengths and elongations to failure of 19.7%. The heat treated ISM sheet performed similarly when compared with the ISM plate. The average properties of the ISM sheet in the final rolling direction were as follows: yield strength 950 MPa, ultimate tensile strength 1033 MPa, and an elongation to failure of 24.6%. The average properties for the ISM sheet in the transverse direction were as follows: yield strength 1039 MPa, ultimate tensile strength was 1087 MPa, and an elongation to failure of 23.1%.

References

1.   Matthew J. Donachie, Jr.; *Titanium: A Technical Guide*, (ASM International, Metals Park, OH 44073, 1988)

2.   R. Boyer, G. Welsch, E.W. Collings; *Materials Properties Handbook: Titanium Alloys*, (ASM, Metals Park, OH, 1994)

3.   Christopher Leyens, Manfred Peters; *Titanium and Titanium Alloys: Fundamentals and Applications*, (WILEY-VCH, Köln, Germany, 2003)

4.   S. Tamirisa, R.B. Bhat, J.S. Tiley, D.B. Miracle, "Grain refinement of cast titanium alloys via trace boron addition." Scripta Materialia 53 (2005): 1421-1426

5.   Howard E. Boyer, Timothy L. Gall; (*Metals Handbook*, ASM, Metals Park, Ohio 44073, 1985)

6.   Matthew J. Donachie, Jr.; (*Titanium and Titanium Alloys*, ASM, Metals Park, OH 44073, 1982)

Effect of Al Content on Phase Constitution and Tensile Properties of Ti-13mass%Cr-1mass%Fe-Al Alloys

Michiharu Ogawa[1], Tetsuya Shimizu[1], Toshiharu Noda[1] and Masahiko Ikeda[2]

[1]Research & Development Lab., Daido Steel Co., Ltd., Nagoya 457-8545, Japan
[2]Department of Mats. Sci. and Eng., Faculty of Eng., Kansai University, Suita 564-8680, Japan

Keywords: Titanium-chromium-iron-aluminum alloy, electrical resistivity, tensile strength, athermal ω, isothermal ω

Abstract

The influence of Al content on phase constitution and tensile properties of Ti-13mass%Cr-1mass%Fe-Al alloys was investigated by measurement of electrical resistivity and Vickers hardness, X-ray diffractometry (XRD) and tensile test. For all solution treated and quenched (STQed) specimens, only β reflection was detected by XRD. Resistivity at room and liquid nitrogen temperatures increased monotonically with Al content. Vickers hardness kept almost the same value as STQed specimen up to 3.0mass%Al and then increase with Al content. In the isochronal heat treatment, two minima of resistivity ratio (ρ_{LN}/ρ_{RT}) were observed at 623K and 823K, respectively. The former is due to the precipitation of isothermal ω phase and the latter is α precipitation. The temperature of resistivity minimum attributing to ω phase precipitation gradually increases with Al content. In the alloys with 4.5mass%Al and more, the minimum of resistivity associated with ω precipitation disappeared and the minimum caused by α precipitation remained. It is considered that Al content above 4.5msss% was suppressed isothermal ω precipitation and enhanced α precipitation. In STQed alloys, tensile strength increased slightly with Al content and elongation at fracture kept almost constant value. Ti-13mass%Cr-1mass%Fe-Al alloys in STQed state without isothermal ω precipitation had the comparable tensile properties to Ti-22mass%V-4mass%Al, one of the most popular β titanium alloy.

Introduction

β titanium alloys have high strength and excellent cold workability, so that the alloys have been often used for the materials of sport applications. However, it is necessary to lower the price of the alloys in order to accelerate the expansion of applications. It is one of the effective ways to substitute chromium and iron for high-cost alloying elements like vanadium as β stabilizer [1, 2].

So far, Ti-Cr-Fe system alloys have been investigated by present authors. Ti-13mass%Cr-1mass%Fe alloy is one of the well-balanced alloys in the system [3]. However, it has a high possibility that the alloy become brittle by isothermal ω precipitates.

179

Generally, it is well-known that Al addition to β titanium alloys suppress isothermal ω precipitation. Therefore, it is important to investigate the influence of Al on the properties of Ti-Cr-Fe system.

In this study, the effect of Al content on phase constitution and tensile properties of Ti-13mass%Cr-1mass%Fe-Al alloys was investigated by measurement of electrical resistivity and Vickers hardness, X-ray diffractometry and tensile test.

Experimental Procedure

Experimental alloys were prepared by a levitation furnace using sponge titanium, chromium and ferro-chromium alloy as raw materials. The roughly 2 kg ingots obtained were hot-forged about 1050 K into about 20mm diameter round bars. The alloy marks and the chemical compositions are shown in **Table 1**. The alloys for tensile test were prepared by a levitation furnace of 10 kg ingots. The chemical compositions of these alloys were also shown in **Table 1**. All specimens were prepared from the hot-forged round bars by cutting, grinding and polishing. Once they were sealed into quartz tubes, all specimens were solution-treated at 1173 K for 3.6 ks and then quenched in water by immersing the tubes in water and immediately breaking the tubes. This solution treatment will hereafter be abbreviated as STQ. **Figure 1** shows schematic diagram of isochronal heat treatment. STQed specimens for aging were isothermally aged by 50 K/3.6ks, i.e. aged for 3.6 ks at temperature which was elevated intervals of 50 K up to the solution treatment temperature. The electrical resistivity at room and liquid nitrogen temperatures (ρ_{RT} and ρ_{LN}) and Vickers hardness (HV) at room temperature were measured. The phase constitution was identified by X-ray diffractometry (XRD). The tensile test samples were also solution-treated at 1173 K for 3.6 ks and then quenched in water. Tensile tests were performed by 0.17 mm/s in cross head speed at room temperature.

Table 1 Alloy marks and chemical compositions of alloys used in this study

Item	Alloy mark	Chemical composition (mass%)				
		Cr	Fe	Al	N	O
Electrical resistivity Vickers hardness	0Al	13.7	1.39	0.02	0.005	0.047
	1.5Al	13.0	1.49	1.57	0.005	0.048
	3.0Al	13.1	1.25	3.12	0.010	0.058
	4.5Al	13.0	1.22	4.62	<0.005	0.051
	6.0Al	13.0	1.24	6.15	<0.005	0.045
Tensile test	0Al	12.6	1.21	<0.01	0.008	0.074
	3.0Al	12.7	1.19	2.98	0.007	0.072
	4.5Al	12.6	1.18	4.47	0.006	0.063
	6.0Al	13.1	1.37	6.17	0.005	0.062

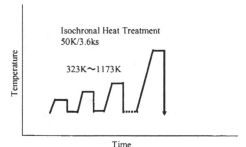

Fig.1 Schematic diagram of isochronal heat treatment.

Results and Discussion

STQed and Isochronal Heat Treatment

Figure 2 shows X-ray diffraction profiles of 0Al to 6.0Al alloys in STQed state. In all STQed specimens, XRD revealed only the β reflection. **Figure 3** shows changes in resistivity (ρ_{RT} and ρ_{LN}), resistivity ratio (ρ_{LN}/ρ_{RT}) and Vickers hardness (HV) with Al content. Resistivity at room and liquid nitrogen temperatures increased monotonically with Al content. These increases were due to the dissolution of Al in matrix [4]. The resistivity at liquid nitrogen temperature was higher than that at room temperature in all alloys. The resistivity ratio increased with increasing Al content up to 3.0mass% and then decreased with increasing Al content. The change of resistivity ratio is due to lowering starting temperature of athermal ω formation with increasing in Al content [4]. Vickers hardness kept almost same value as the STQed specimen up to 3.0mass%, and then increase with Al content. It is considered that the hardness change causes balance between the decrease of volume fraction of athermal ω with increasing Al content and solution hardening by Al addition [5]. The similar changes of resistivity ratio and Vickers hardness were already observed in Ti-V-Al and Ti-Mo-Al alloys [6].

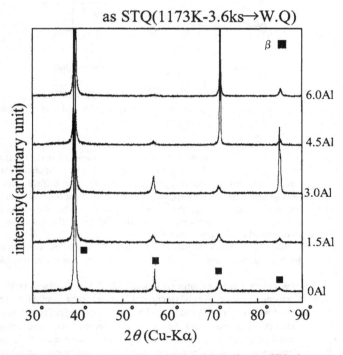

Fig.2 X-ray diffraction profiles of 0Al to 6.0Al alloys in STQed state.

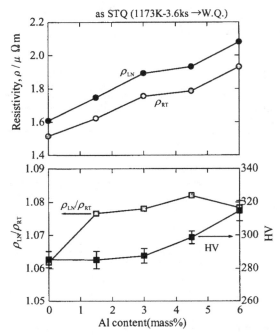

Fig.3 Changes in resistivity (ρ_{RT} and ρ_{LN}), resistivity ratio (ρ_{LN}/ρ_{RT}), and Vickers hardness (HV) with Al content.

Figure 4 shows selected area electron diffraction (SAED) patterns of 0Al, 3.0Al and 6.0Al alloys quenched from 1173K. These diffractions were obtained at room temperature. In 0Al and 3.0Al, β reflection and weak diffuse scattering were observed. The diffuse scatterings are attributed to the present of athermal ω. In 6.0Al, only β reflections were observed. From change of the SAED pattern, it is considered that the volume fraction of athermal ω decrease with increasing of Al content at room temperature. Thus, the change in the resistivity ratio and Vickers hardness can be linked to the change in the SAED pattern.

Figure 5 shows changes in X-ray diffraction profiles of 0Al alloy with increasing in temperature of isochronal heat treatment. The reflection of isothermal ω phase was identified at 623K and 673K.

Figure 6 shows changes in resistivity ratio (ρ_{LN}/ρ_{RT}) and Vickers hardness (HV) of 0Al to 6.0Al alloys with increasing in temperature of isochronal heat treatment. In 0Al, two minima of resistivity ratio (ρ_{LN}/ρ_{RT}) were observed at 623K and 823K, respectively. The former is due to the precipitation of isothermal ω phase and the latter is α precipitation. The temperature of resistivity minimum attributing to ω phase precipitation gradually increases with Al content. In 4.5Al and 6.0Al, the minimum of resistivity associated with ω precipitation disappeared and the minimum caused by α precipitation remained. It is considered that Al content above 4.5msss% was suppressed isothermal ω precipitation and enhanced α precipitation.

182

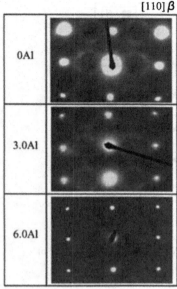

[110] β

0Al

3.0Al

6.0Al

Fig.4 Comparison of selected area
electron diffraction patterns of 0Al,
3.0Al and 6.0Al alloys quenched
form 1173K.

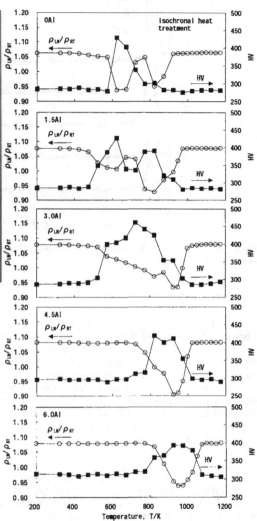

Fig.6 Changes in resistivity ratio (ρ_{LN}/ρ_{RT}) and
Vickers hardness (HV) of 0Al to 6.0Al alloys
with increase in temperature of isochronal heat
treatment.

183

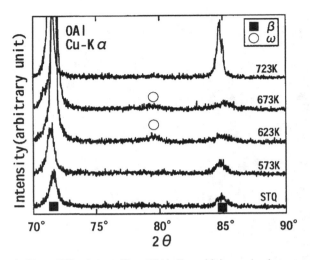

Fig.5 Changes in X-ray diffraction profiles of 0Al alloy with increasing in temperature of isochronal heat treatment.

Tensile Properties

In all STQed alloys for tensile test, reflections from β phase were only identified by XRD. **Figure 7** shows changes in tensile strength (σ_B), 0.2% proof stress ($\sigma_{0.2}$), elongation δ and reduction in area (φ) with increasing Al content. σ_B and $\sigma_{0.2}$ kept almost constant up to 4.5Al. In the alloys, σ_B and $\sigma_{0.2}$ were about 930MPa and 900 MPa, respectively. The δ and φ kept almost constant up to 4.5Al which δ and φ were about 20% and 60%. In 6.0Al, σ_B and $\sigma_{0.2}$ increased slightly, while δ and φ decreased slightly. The increases of both strengths are due to solution hardening by Al. Though the cause of decrease in δ and φ is not clear, there is possibility that deformation mode changes around 4.5Al. These alloys in STQed state without isothermal ω precipitation show good balance between strength and ductility. For example, the σ_B, δ and φ of Ti-22%V-4%Al, a commercial β titanium alloy, are about 760 MPa, 22% and 65% [7].

Conclusions

The influence of Al content on phase constitution and tensile properties of Ti-13mass%Cr-1mass%Fe-Al alloys was investigated by measurement of electrical resistivity and Vickers hardness, X-ray diffractometry (XRD) and tensile test.

For all solution treated and quenched (STQed) specimens, only β reflection was detected by XRD. Resistivity at room and liquid nitrogen temperatures increased monotonically with Al content. Vickers hardness kept almost the same value as STQed specimen up to 3.0mass%Al and then increase with Al content. In the isochronal heat treatment, two minima of resistivity ratio (ρ_{LN}/ρ_{RT}) were observed at 623K and 823K, respectively. The former is due to the precipitation of isothermal ω phase and the latter is α precipitation. The temperature of resistivity minimum attributing to ω phase precipitation gradually increases with Al content. In the alloys with 4.5mass%Al and more, the minimum of resistivity associated with ω precipitation disappeared and the minimum caused by α precipitation remained. It is considered that Al content above 4.5msss% was suppressed isothermal ω precipitation and enhanced α precipitation. In STQed alloys, tensile strength increased slightly with Al content and elongation at fracture kept almost constant value. Ti-13mass%Cr-1mass%Fe-Al alloys in STQed state without isothermal ω precipitation had the comparable tensile properties to Ti-22mass%V-4mass%Al, one of the most popular β titanium alloy.

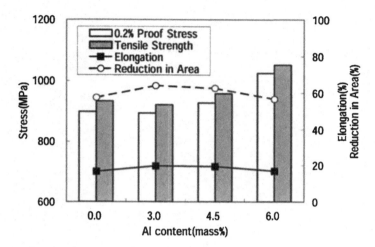

Fig.7 Changes in tensile strength (σ_B), 0.2% proof stress ($\sigma_{0.2}$), elongation (δ) and reduction in area (ϕ) with increase on Al content.

References

1.   M. Ikeda, S. Komatsu, M. Ueda, T. Imose, and K. Inoue, "Influence of Aluminum Addition on Tensile Properties and Aging Behavior of Ti-Fe-Cr-Al Alloys," The Fourth Pacific Rim International Conference on Advanced Materials and Processing (PRICM4), ed. by S. Hanada, Z. Zhong, S. W. Ham, and R. N. Wright, (JIM, Sendai, 2001), 213-216.
2.   M. Ikeda, S. Komatsu, M. Ueda, and A. Suzuki, "The Effect of Cooling Rate from Solution Treatment Temperature on Phase Constitution and Tensile Properties of T-4.3Fe-7.1Cr-3.0Al Alloy," Mater. Trans., 45 (2004), 1566-1570.
3.   M. Ikeda, S. Komatsu, M. Ueda, and K. Inoue, "The effect of aluminum content on tensile properties and aging behavior of Ti-Fe-Cr-Al alloys," CAMP-ISIJ, 14 (2001), 1340.
4.   M. Ikeda, S. Komatsu, T. Sugimoto, and K. Kamei,"Influence of Al addition on electrical resistivity and phase constitution in quenched Ti-20mass%V alloy," J. J. I. L. M., 42 (1992), 622-626.
5.   M. Ikeda, S. Komatsu, T. Sugimoto, K. Kamei, and K. Inoue, "Influence of Al addition on Resistivity and Phase Constitution of Quenched Ti-5Fe Alloy," Tetsu-to-Hagane, 80 (1994), 866-870.
6.   J. C. Williams, B. S. Hickman, and D. H. Leslie, " The Effect of Ternary Additions on the Decomposition of Metastable Beta-Phase Titanium Alloys," Met. Trans., 2 (1971), 477-484.
7.   S. Fukui, Y. Ohtakara, and A. Suzuki, "Properties of a Beta Type New Titanium Alloy Ti-22V-4Al," Electric Furnace Steel, 57 (1986), 303-317.

Innovations in Titanium Technology

Microstructure and Properties I

MICROSTRUCTURE AND MECHANICAL PROPERTIES OF Ti-6Al-4V INVESTMENT CASTINGS

Ibrahim Ucok[1], Hao Dong[1], Kevin L. Klug[1], Lawrence S. Kramer[1], Mehmet N. Gungor[1] and Wm. Troy Tack[1]

[1] Concurrent Technologies Corporation
100 CTC Drive; Johnstown, PA 15904, USA

Keywords: Investment Casting, Ti-6Al-4V Alloy, Weld-repair, Surface Roughness, Mechanical Properties, Fatigue

Abstract

The microstructure and mechanical properties of investment cast Ti-6Al-4V have been studied as a function of casting thickness, weld-repair, heat treatment and surface finish. Standard metallography and scanning electron microscopy fracture analysis techniques were utilized to characterize the material specimens. Tensile and fatigue tests were performed on specimens extracted from cast components as well as on separately cast bars and plates. The results were analyzed as a function of section thickness, weld repair, heat treatment and surface roughness.

Introduction

Given the high price and environmental impacts of energy consumption, it is apparent that weight reduction in all transportation vehicles will have a significant effect on both energy savings and the environment. These conditions make titanium (Ti) alloys very attractive because they have high specific strength in both wrought and cast forms, exhibit very good damage tolerance properties and possess excellent corrosion resistance. Due to these attributes, Ti alloys are widely used for both military and commercial aerospace applications and more recently have been used in military ground vehicles and weapon systems [1–6]. However, Ti alloys are expensive, mainly due to their high affinity for oxygen and hence complexity of extraction processes. Melting and alloying, as well as loses during secondary process operations such as forging, rolling and casting, also contribute to high cost. Therefore, there have been great efforts to reduce the cost of Ti materials and products in recent years [7]. There is especially great emphasis on reducing the extraction cost of Ti alloys and numerous technologies are still in the developmental stage [7]. In addition, there is a strong market demand for using traditional and state-of-the-art casting processes to produce larger, monolithic structures and more near-net-shape (NNS) cast parts. These approaches will further reduce the part count and therefore, the amount of tooling and floor space required for assembly, reduce the machining time, and decrease the assembly time; all key elements in reducing manufacturing costs [1–4]. Coupling these with the excellent corrosion resistance and high specific strength, monolithic Ti castings become economical when life cycle cost is considered. The use of a large percentage of revert and plasma arc consolidation of electrodes for VAR melting prior to casting enables some additional cost reduction and has been applied to non-aerospace applications [1–6].

In this study, two castings that were produced for the M777 Lightweight Howitzer (LWH) using large amounts of revert, e.g., casting scrap, were evaluated. These components were hard-tooled investment castings, designated Spade and Saddle, and were made by two different vendors. The main objective of this paper is to summarize and assess the microstructure and mechanical properties of these relatively large titanium investment castings developed for the LWH application. The mechanical properties determined include tensile and fatigue results from various locations of these castings having different thickness, weld-repaired areas, and surface defects. Figure 1 shows photographs of the investment cast components.

(a) Spade   (b) Saddle

Figure 1. Investment cast components with the longest dimension of 100 cm and weights of (a) 45 kg and (b) 160 kg, respectively [1, 4].

Experimental Work

Casting and Post-casting Processes – For both cast components, wax patterns were produced utilizing hard tools. An investment shell was built around the wax pattern by multiple dipping and drying cycles. Molten Ti-6Al-4V was poured into fired shell molds. After cooling and mold removal, standard processing steps such as cleaning, hot isostatic pressing (HIPing), chemical milling, fluorescent penetrant inspection, radiography, weld-repair, and heat treatment procedures were applied separately to both castings. Specimens attached to the cast components were also subjected to the same procedures. HIPing was conducted at 900°C in pure argon under 103 MPa pressure for two hours. Saddle was subjected to a stress-relief heat treatment at 566°C for four hours. Saddle attached specimens were subjected to (i) a stress relief at 566°C for four hours or (ii) an anneal at 732°C for two hours. Spade and its attached specimens were subjected to annealing at 843°C for two hours. Chemical analyses were conducted on samples attached to the castings. Metallographic inspection to verify alpha case removal was conducted on a sample taken from the thickest section of each cast part.

Tensile Testing – Tensile tests were performed at room temperature in accordance with ASTM E 8 [8]. Specimens representing different section thickness, weld-repair areas and as-cast surface roughness were taken from various locations of each cast component. Figure 2 shows some of the specimen locations marked on the cast components.

Fatigue Testing – Fatigue tests were performed at room temperature at R-ratios of 0.1 and –1.0 in accordance with ASTM E 466 [9] using specimens with both circular and rectangular cross sections from various locations and thicknesses of Saddle and Spade; see Figure 2 for locations. Notched (K_t = 3.0) specimens from Saddle were also tested using the same procedures. Additionally, specimens with rectangular cross-section were tested with as-cast + chemically milled surface. These specimens did not comply with the surface requirements of ASTM E 466 but were tested to compare their behavior to that of polished specimens. Other specimens contained weld-repair in the gage section.

Microstructure Examination and Fractography – The microstructural features of specimens taken from various locations of both castings were revealed by grinding, polishing and etching (Kroll's reagent) followed by optical microscopy. The fracture surfaces of selected tensile and fatigue specimens were examined by scanning electron microscopy (SEM) to reveal fracture features such as crack initiation location.

190

Figure 2. Component and segments with mechanical test specimen locations marked.

Results and Discussion

<u>Casting and Post-casting Processes</u> – The chemical compositions of cast Saddle and Spade are given in Table I along with ASTM B 367 requirements [10]. Both castings met the ASTM specification compositional requirements. As seen from the table, Saddle has slightly higher oxygen (0.20 wt% O) compared to 0.15 wt% O of Spade. The thermal history of both castings and attached specimens are summarized in Table II. Saddle was heat treated at a significantly lower temperature of 566°C for four hours compared to the 843°C, two hour heat treatment of Spade. Note that some of the Saddle attached specimens were also annealed at 732°C for two hours to enable comparison between stress relief and annealing heat treatments.

Table I. Chemical Composition of Investment Cast Components Compared to ASTM B 367 Grade C-5 Requirements [10], in wt%

Component I.D. and Heat No	C	O	N	H	Al	V	Fe	Y	Other, total
Saddle, AN 9468 JN 37155	0.02	0.20	0.02	0.01	5.96	3.90	0.18	NA	NA
Spade, 30555-001	0.03	0.15	0.04	0.002	6.22	3.99	0.20	<0.0010	NA
ASTM B 367 C-5[1]	0.04	0.25	0.05	0.015	5.5/6.75	3.5/4.5	0.40	NA	0.4

[1] Single values are maxima.

Table II. Heat Treatments Applied to Investment Saddle and Spade and their Attached Specimens

Component I.D. and Heat Number	Description	Processing Conditions
Saddle, AN 9468 JN 37155	Casting	HIPed at 900°C, 103MPa argon for two hours. Stress-relieved at 566°C for four hours after weld repair.
	Attached specimens	
	Attached specimens	HIPed at 900°C, 103MPa argon for two hours. Annealed at 732°C for two hours after weld repair.
Spade, 30555-001	Casting	HIPed at 900°C, 103MPa argon for two hours. Annealed at 843°C for two hours after weld repair.
	Attached specimens	

<u>Tensile</u> – The tensile properties measured from bars attached to the castings (Saddle and Spade) are given in Table III. Similarly, the tensile properties of plates cast with Saddle and Spade are summarized in Table IV. Stress-relieved (566°C for four hours) specimens attached to Saddle exhibited higher UTS and YS values than those of annealed (732°C for two hours) specimens. Specimens attached to Spade, which were annealed at 843°C for two hours, exhibited the lowest strength values among all attached bars. It is worthwhile to note that Spade had a lower oxygen level compared to Saddle in addition to high heat treatment temperature. Elongation values appear similar for all three types of attached bars. However, reduction of area values varied between 23.4% and 26.9%. Note that Saddle attached bars complied with the ASTM B 367 C-5

191

requirements, whereas Spade attached bars, which were heat treated at a higher temperature than Saddle bars were, were slightly below the recommended strength values.

Table III. Tensile Properties of 6.4mm-diameter Attached Test Bars from Saddle and Spade (Mean of 3) Compared to ASTM B 367 C-5 Separately Cast Bars

Attached to	Heat Treatment	UTS (MPa)	YS (MPa)	el. (%)	RA (%)
Saddle	Stress relieved, 566°C x 4h	1000	900	14.6	24.8
	Annealed, 732°C x 2h	965	832	14.6	23.4
Spade	Annealed, 843°C x 2h	888	797	14.7	26.9
ASTM B 367 C-5	NA	896	827	6.0	NA

Table IV. Tensile Properties of Plates Cast with Saddle and Spade (Mean of a Minimum of 3)

Thickness (mm) & I.D.	Heat Treatment	UTS (MPa)	YS (MPa)	el. (%)	RA (%)
13, Saddle	Stress relieved, 566°C x 4h	936	855	10.7	21.0
	Annealed, 732°C x 2h	912	821	13.0	22.0
	Welded + stress relieved, 566°C x 4h	1071	974	8.0	12.7
25, Saddle	Stress relieved, 566°C x 4h	915	839	8.3	16.7
	Annealed, 732°C x 2h	912	823	6.3	15.3
13, Spade	Annealed, 843°C x 2h	896	814	11.7	17.7
25, Spade	Annealed, 843°C x 2h	896	821	7.7	14.7

All 13mm- and 25mm-thick plates cast with Saddle and heat treated at 566°C x 4h or 732°C x 2h (Table IV) exhibited both lower strength and ductility values compared to those of Saddle-attached bars heat treated at the same temperatures (Table III). 13mm-thick plate in the annealed (732°C x 2h) condition exhibited the highest ductility (elongation and reduction of area of 13% and 22%) values among the thickness and heat treatment conditions compared for the plates. Note that 25mm-thick plate cast with Saddle annealed at 732°C x 2h exhibited the lowest elongation and reduction of area values of 6.3% and 15.3%, respectively, although there was no significant difference between strength values of UTS and YS of 915 MPa and 839 MPa, respectively, for the plates heat treated at 566°C x 4h, and UTS and YS of 912 MPa and 823 MPa, respectively, for the plates heat treated at 732°C x 2h. The observed drop in ductility values may be attributed to coarsened microstructures as a result of increased cast section thickness and increased heat treatment temperature. It seems that the section thickness effect is more pronounced than the heat treatment effect in the temperature range used. By observing Table IV, one can see that the drop in ductility values is larger when moved from 13mm- to 25mm-thick plates for the same heat treatment. This is true for the plates cast with Saddle and Spade. When thickness is kept constant and heat treatment is changed, as in the case of 25mm-thick plates, a smaller drop observed between ductility values, for instance, when moved from plate heat treated at 566°C x 4h (with elongation and reduction of area of 8.3% and 16.6%) to plate heat treated at 732°C x 2h (with elongation and reduction of area of 6.3% and 15.3%).

Welded + stress relieved (at 566°C x 4h), 13mm-thick plate, exhibited the highest UTS and YS of 1071 MPa and 974 MPa, respectively, along with 8% elongation and 12.7% reduction of area. The ductility values for the welded plate were lower compared to the non-welded 13mm-thick plate with the same heat treatment.

Table V and Figure 3 summarize the mean tensile properties of Saddle and Spade castings as a function of cast section thickness. Each data point is the mean value of a minimum of three values. For Saddle, the mean UTS and YS values exhibit a slight drop when the cast section thickness increases from 5mm to 20mm. Elongation values also exhibit a decrease when the cast section thickness increases from 10mm to 30mm. Reduction of area values are somewhat scattered. For Spade (Table V), mean strength values seem to stay in the same range, for UTS from 862 to 880 MPa and YS from 767 to 795 MPa, respectively for all cast section thicknesses considered. However, mean elongation (from 12.0% to 9.7%) and reduction of area (from 20.6 to 15.0%) values exhibit a decline with increasing casting thickness from 5mm to 28mm.

Table V. Mean Tensile Properties of Ti-6Al-4V Cast Components (Mean of 3)

Component I.D.	Heat Treatment	Mean Section Thickness (mm)	UTS (MPa)	YS (MPa)	el (%)	RA (%)
Saddle	Stress relieved, 566°C x 4h	5.0	951	856	9.9	16.3
		10.0	940	853	10.3	21.0
		20.0	911	833	9.7	21.8
		25.0	927	853	7.7	12.5
		30.0	912	839	7.7	21.0
Spade	Annealed, 843°C x 2h	5.2	862	767	12.0	20.6
		8.1	879	790	11.2	16.9
		10.3	880	795	10.4	14.6
		15.2	865	793	8.0	13.5
		27.9	869	786	9.7	15.0

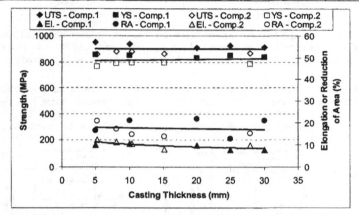

Figure 3. Mean tensile properties of Saddle and Spade components depending on cast section thickness (each reported value is the mean of a minimum of three data points).

Fatigue – Figure 4a shows the stress versus number of cycles (S-N$_f$) curve developed per the recommended practice of ASTM E 739 [11] using specimens from both Saddle and Spade. Specimens with rectangular cross-section taken from various cast thickness sections (between 3.6mm and 20mm) were tested using R of –1.0. 99% confidence limits for the S-N$_f$ curve are also shown on the plot. Note that the legends for Saddle and Spade are identified on the plot. The data used in the curve fit are shown by solid markers whereas the data that are not included in the curve fit (due to failure location being outside the gage section) are indicated by empty markers and they were shown in the plots for reference. Specimens that did not fail at 10^7 cycles of life are called run-outs and were excluded from the regression analysis. The curve fit equation, standard error and R^2 value for the curve fit are shown in the figure as well. Run-outs on the S-N$_f$ curve are observed between 241 MPa and 276 MPa marked with arrows. These values are similar to those reported in the literature [12–14].

In Figure 4b, a comparison is shown between S-N$_f$ curves of rectangular cross-section specimens machined from the 10mm thick section of Saddle and 13mm-thick weld-repaired plates (all in the stress-relieved condition). Although one set of specimens was machined from the casting

193

and the other from cast plate, this was the best comparison available of cast metal and weld-repaired metal of a comparable thickness. Tensile results showed that tensile specimens machined from the casting and the plate were found to have equivalent properties, although the weld-repaired cast plate displayed significantly higher strength than the non-welded cast plate.

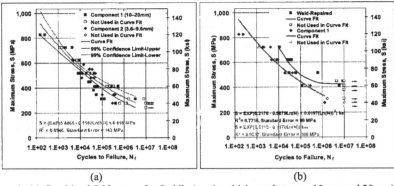

(a)   (b)

Figure 4. (a) Combined S-N$_f$ curve for Saddle (casting thickness between 10mm and 20mm) and Spade (casting thickness between 3.6mm and 9.6mm). 99% confidence limits are also shown. (b) Comparison of fatigue curves for specimens taken from Saddle vs. weld-repaired plate, R = –1.0.

Indeed, the weld-repaired S/N$_f$ curve lies above that of the non-weld-repaired specimens. For medium stress levels, several fatigue data points overlap. However, there is a significant difference of about 138 MPa in the endurance limits between the two variants. This difference was attributed to the significantly higher tensile strength values of the weld-repaired plate.

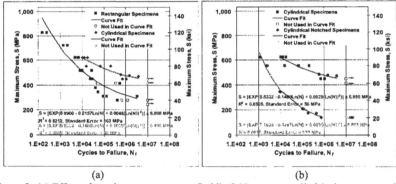

(a)   (b)

Figure 5. (a) Effect of specimen geometry on Saddle S-N$_f$ curves: cylindrical vs. rectangular, and (b) comparison of Saddle S-N$_f$ curves developed using cylindrical smooth (K$_t$ = 1.0) and notched (K$_t$ = 3.0) specimens (all R = –1.0).

Most of the fatigue specimens taken from Saddle had a rectangular cross-section because extraction of cylindrical specimens from thin sections was not possible. The specimens with rectangular cross-section were machined from the center of the section or used the full thickness of the section in the case of the thinnest areas of the casting. Additional specimens with circular cross-section were taken from the 20mm thick location to facilitate comparison with the rectangular-cross-section specimens from the same location. Figure 5a shows the effect of

194

specimen geometry on fatigue behavior of specimens from 20mm thick section with two different geometries tested at R of –1.0. Clearly, there is a significant difference between the two curves. Although both group of specimens had smooth surfaces, a nominal K_t of 1.0, and possessed nominal standard surface finishes of 0.20 µm R_a, specimens with rectangular cross-section might not have a truly uniform K_t across their faces. The edge effects during loading/unloading cycles might have led to premature failure in specimens with rectangular cross-section. Specimens with circular cross-section did not suffer from this effect.

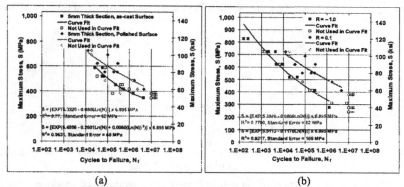

(a)   (b)

Figure 6. (a) Effect of surface roughness on Saddle S-N_f curves: as-cast (R_a = 2.19µm) and polished surface (R_a = 0.20µm) compared for R = 0.1, and (b) Saddle S-N_f curves developed using two different R values of –1.0 and 0.1 (all rectangular specimens).

The results of smooth and notched cylindrical fatigue specimens machined from 20mm thick sections of Saddle are presented in Figure 5b. The stress concentration factor (K_t) for smooth specimens and notched specimens are 1.0 and 3.0, respectively. As expected, the smooth specimens greatly outperformed the notched specimens, although the curves are closer to each other at higher stress levels that correspond to the low-cycle fatigue (LCF) regime. Clearly, severe stress concentrations are not desired in titanium castings because they would lead to premature failure.

Figure 6a compares S-N_f curves of Saddle specimens with rectangular cross-section tested in both the as-cast and machined/polished condition. Polished specimens have a standard surface finish of 0.20µm R_a. As-cast + chemically milled specimens exhibited surface finishes between 1.57µm and 2.77µm with the mean measured surface finish of 2.19µm±0.35µm. The mean as-cast surface roughness for each specimen is about an order of magnitude larger than the standard fatigue specimen finish. The specimens polished to R_a of 0.20µm show a clear advantage for increased fatigue life, with the S/N_f curves separated by about an order of magnitude of cycles of life.

A comparison of R-ratios (0.1 vs. –1.0) is shown by the fatigue curves in Figure 6b. Although these rectangular specimens are from different thickness regions of Saddle, they show very similar tensile properties as shown in Table V and therefore they should exhibit similar fatigue behavior. Note that fatigue data from some specimens are not included in the regression analysis and curves because these specimens failed outside the gage. The fatigue curve for R = 0.1 exhibits a significant improvement in fatigue strength (and life) compared to R = –1.0. It is interesting to note that the difference in fatigue life for the same strength in the S/N_f curves of the two R-ratios tested is more than one order of magnitude.

195

Microstructure – Figure 7 shows microstructures from 5mm, 10mm and 25mm thick sections of Saddle. Spade exhibited similar microstructures. The typical microstructure consisted of coarse alpha plates separated by intergranular beta, primary alpha at prior beta grain boundaries, and alpha colonies. Both grain boundary alpha and alpha colonies have been shown to reduce fatigue resistance by causing fatigue crack initiation [12]. However, they increase the resistance to fatigue crack growth rate and improve fracture toughness [12–15]. Quantitative measurements on plate specimens cast with Spade revealed that the prior beta grain size increased with increasing section thickness. Measured volume fraction of alpha phase varied between 66% and 71%.

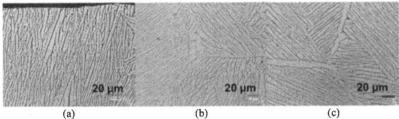

(a)   (b)   (c)

Figure 7. Typical microstructures of Saddle from locations having different section thickness: (a) 5mm, (b) 10mm, and (c) 25mm.

Fractography – Optical microscopy on polished and etched tensile tested specimens (Spade) showed that the fracture surface traversed alpha plates at many different angles; alpha plates oriented parallel, perpendicular and at other angles to the fracture surface are shown in Figure 8a. This is in agreement with the literature. In lamellar microstructures of Ti-6Al-4V castings, changes in crack propagation direction occur when a crack reaches the end of an alpha colony and changes direction depending on the slip direction in the next alpha colony [12, 15], generating the jagged morphology. Cracks also propagate in the alpha phase along the prior beta boundaries. Note that the secondary cracks away from fracture surface, shown in Figure 8b and 8, developed at alpha colony boundary and grain boundary alpha, respectively, as explained in [12–15].

SEM fractography on selected specimens also revealed the jagged morphology of fracture surface, Figure 9. This morphology develops due to changes in crack propagation direction as a result of microstructure differences [12, 15]. Crack initiation sites are at the surface of both specimens and marked by arrows on the fractographs. Figure 10 shows a SEM image of a Spade specimen revealing fatigue striations that appear to be of a finer scale than the alpha plate spacing seen on the "polished and etched" side surface of the specimen.

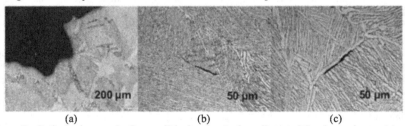

(a)   (b)   (c)

Figure 8. Optical micrographs from polished and etched tensile tested Spade specimen: (a) part of the primary fracture path, and secondary cracks (away from fracture surface) at (b) alpha colony boundary and (c) grain boundary alpha.

196

(a)   (b)

Figure 9. SEM fractographs of broken fatigue specimens from Saddle tested at maximum stress
of (a) 483 MPa with cylindrical and (b) 517 MPa with rectangular geometry, R = −1.0.

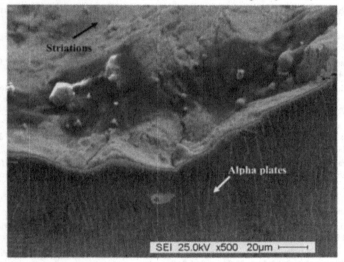

Figure 10. SEM tilted view image of a fatigue specimen from plate cast with Spade tested at 550
MPa maximum stress showing microstructure and fatigue striations.

Summary and Conclusions

In this study, NNS investment cast Ti-6Al-4V components produced using large amounts of
revert were evaluated. Microstructure characterization, tensile and fatigue testing were
conducted on specimens in stress relieved or annealed conditions. Based on experimental results
the following is concluded.

- Microstructures were similar to those reported in the literature.
- Mean tensile properties met ASTM B367 Grade C-5 requirements.
- Fatigue properties were similar to those reported in the literature for titanium investment
  castings. Cylindrical specimens outperformed rectangular specimens in fatigue tests.
  Smooth specimens exhibited run-outs at 400–450 MPa whereas notched specimens (K_t of
  3.0) exhibited run-outs at 138 MPa. Polished specimens outperformed specimens with

as-cast + chemically milled surfaces in fatigue tests. Fatigue strength values for R = − 1.0 were significantly lower than those of R = 0.1.

Based on mechanical property data, Ti-6Al-4V investment castings made by using large amounts of revert are viable for land based applications.

Acknowledgement

This work was conducted by the Navy Metalworking Center, operated by Concurrent Technologies Corporation under Contract No. N00014-00-C-0544 to the Office of Naval Research as part of the U.S. Navy Manufacturing Technology Program. Approved for public release, distribution is unlimited.

References

1. K. L. Klug, I. Ucok, M. N. Gungor, M. Guclu, L. S. Kramer, Wm. T. Tack, L. Nastac, N. R. Martin and H. Dong, "Near-Net-Shape Manufacturing of Affordable Titanium Components for M777 Lightweight Howitzer" "Near-Net-Shape Manufacturing of Affordable Titanium Components for the M777 Lightweight Howitzer," *JOM*, Vol. 56, No. 11, 2004, 35-41.

2. C. Hatch and R. Nestor, "Military Makeover – The Investment Casting Way," *Modern Casting*, Vol. 93, No. 12, 2003, pp. 30–32.

3. F. Hoerster and J. Boulet, "Investment Castings Pioneer New Applications," *Foundry Management & Technology*, June 2004, pp. 14–17.

4. L. Nastac, M.N. Gungor, I. Ucok, K. L. Klug and Wm. T. Tack, "Advances in Investment Casting of Ti-6Al-4V Alloy: A Review," *IJCMR*, Vol. 19, 2006, No.2, 73-93.

5. M. Guclu, I. Ucok and J. R. Pickens, "Effect of Oxygen Content on Properties of Cast Alloy Ti-6Al-4V," in *Cost-Affordable Titanium*, Edited by F. H. Froes, M. A. Imam and D. Fray, The Minerals, Metals & Materials Society, Warrendale, PA, 2004, pp. 135–143.

6. K L. Klug, M.N. Gungor, I. Ucok, C. Hatch, R. Spencer and R. Lomas, "Affordable Ti-6Al-4V Castings" in *Cost-Affordable Titanium*, Edited by F. H. Froes, M. A. Imam and D. Fray, The Minerals, Metals & Materials Society, Warrendale, PA, 2004, pp. 103–109.

7. E.H. Kraft, "Summary of Emerging Titanium Cost Reduction Technologies," A Study performed for US DOE and Oak Ridge National Laboratory, Subcontract 4000023694, EHK Technologies, Jan. 2004,

8. ASTM E 8, Standard Test Methods for Tension Testing of Metallic Materials, Vol. 03.01, ASTM, W. Conshohocken, PA, 2001, pp. 56-76..

9. ASTM E 466, Standard Practice for Conducting Force Controlled Constant Amplitude Axial Fatigue Tests of Metallic Materials, ASTM, W. Conshohocken, PA, 2000, pp. 493-497.

10. ASTM B 367-93 Standard Specification for Titanium and Titanium Alloy Castings, ASTM, W. Conshohocken, PA, 2000, Vol. 02.04, pp 264-268.

11. ASTM E 739, Standard Practice for Statistical Analysis of Linear or Linearized Stress-Life (S-N) and Strain-Life Fatigue Data, ASTM, W. Conshohocken, PA, 2000, pp. 631-637.

12. D. Eylon, "Fatigue Crack Initiation in Hot Isostatically-Pressed Ti-6Al-4V Castings," *Journal of Materials Science*, Vol. 14, 1979, 1914–1920.

13. D. Eylon and J. A. Hall, "Fatigue Behavior of Beta-Processed Titanium Alloy IMI-685," *Metallurgical Transactions* A, Vol. 8A, 1977, 981–990.

14. D. Eylon, F. H. Froes, and R. W. Gardner, "*Development in Titanium Alloy Casting Technology,*" in *Titanium Technology: Present Status and Future Trends*, Edited by F. H. Froes, D. Eylon, and H. B. Boomberger, Titanium Development Association, Broomfield, CO, 1985, 35–47.

15. A.W. Thompson, "Relationship between Microstructure and Fatigue Properties of Alpha-Beta Titanium Alloys," Proc. Int'l Symposium, TMS Fall Meeting, October 11–15, 1998, Chicago, IL, in *Fatigue Behavior of Titanium Alloys*, Edited by R. R. Boyer, D. Eylon and G. Lütjering, TMS, 1999, pp. 23–30.

DENTAL TITANIUM CASTING AT
BAYLOR COLLEGE OF DENTISTRY - UPDATE

Mari Koike[1], Kwai S. Chan[2] and Toru Okabe[1]

[1]Baylor College of Dentistry, Texas A&M Health Science Center,
Dallas, TX 75246
[2]Southwest Research Institute, San Antonio, TX 78238

Keywords: Titanium, Alloy, Mechanical properties, Dental prostheses

Abstract

In 2001, we presented the results of properties testing relevant to dental applications for experimental binary titanium alloys with Ag, Co, Cr, Cu, Fe, Mn or Pd. Tensile strengths, casting performance, grindability, wear resistance, electrochemical behavior and biocompatibility were examined. During the last 5 years, we have tested more titanium alloys, including some ternary alloys, in search of good candidate alloys for dental prosthetic use. Of particular interest was the effect of Cu at improving the grindability and wear resistance, and the strengths of some alloys. Cu was added to alloys such as Ti-Si and Ti-Al alloys to test its effectiveness. In addition to the above characteristics tested, we also evaluated the fatigue resistance, fracture toughness and CAD/CAM applications of selected alloys. The most recent results of our properties characterization of candidate titanium casting alloys for dental prostheses are summarized.

Introduction

Because of great interest in titanium for dental prostheses, the fabrication technology and engineering to make cast dental appliances have vastly improved during the last 15 years. Dedicated casting machines and mold materials are available to cast titanium for clinical use [1,2]. Increasing concerns about corrosion resistance, biocompatibility and allergic reactions to dental and medical prostheses from other materials have established titanium as a preferable material for these applications. This article provides an up-date to our report published in 2001 on the development of castable, inexpensive titanium alloys with suitable corrosion characteristics and biocompatibility.

Alloy Development

Various noble and precious alloys, and non-precious base metal alloys have traditionally been used for dental and biomedical applications. However, current biomedical alloys must be made of materials with high corrosion resistance and biocompatibility with a negligible release of ions [3], which results in an absence of allergic reactions. One direction for developing biomedical titanium alloys is to formulate β titanium alloys since they offer an excellent combination of high strength and low modulus. Sometimes these alloys are used after heat treatment to improve their mechanical properties. Some alloying elements for these β alloys are Nb, Ta, Mo, and Zr [4,5]. However, these metals, particularly Mo and Ta, have a higher melting point than titanium, so titanium alloys with these elements are not easy to melt and cast. Many of these metals are also expensive. Our goal has been to develop inexpensive, biocompatible titanium alloys that are easy to form into fixed and removable dental prostheses by casting without any heat treatments. We have made experimental alloys using low-melting, simple metals (e.g., Al, Si, Sn) and

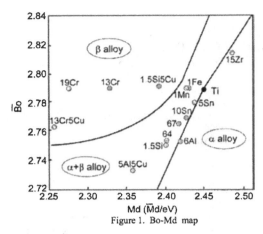

Figure 1. Bo-Md map

transition metals (e.g., Fe, Co, Cr, Cu, Hf, Mn, Zr). The melting points of Hf and Zr are much higher (especially for Hf) than that of Ti, but with increasing Hf or Zr, the binary Ti-Hf and Ti-Zr alloys form congruent alloys in some compositional ranges with a lower melting temperature than that of Ti [6].

Using the Bo-Md map developed by Morinaga et al. [7] , the possible phase-stabilized areas of our experimental alloys with those of Ti-6Al-4V (64) and pure Ti are shown in Fig. 1. Bo is the covalent bond strength between titanium and the alloying elements while Md is the metal d-orbital energy level that correlates with electronegativity and the radius of the elements. Many of our alloys fall in the α+β alloy region of the graph. The alloys with a group number value [8] higher than 4.1, such as 13% Cr and 19% Cr, are located in the β phase area.

Mechanical Properties

Alloys for dental use must meet certain requirements. In addition to acceptable corrosion behavior and biocompatibility, they must exhibit at least similar or better performance compared to the presently used dental casting alloys. The mechanical properties of these metals are specified by the American National Standards Institute/American Dental Association (ANSI/ADA) Specifications for dental casting alloys (Specification No. 5 for alloys in gold and Pt groups, and No. 14 for base metals). In each alloy category, the range for each property is large, indicating some leeway in the mechanical properties when fabricating dental prostheses.

Following our philosophy about the direction that titanium alloy development should take, we made various alloys by melting sponge titanium with alloying constituents in an arc-melting furnace. The tensile strengths for selected cast experimental alloys and industrial titanium alloys are shown in Fig. 2. The values shown were obtained at Baylor College of Dentistry using CP Ti cast in a MgO-based investment material in a centrifugal casting machine [9].

Information for the tested metals shown in Fig. 2 is given in Table 1. In the evaluation of industrial alloys, cast pure titanium exhibited the lowest strength, whereas cast Ti-6Al-4V has the highest strength. The range of strength of the cast experimental alloys was similar for the cast industrial alloys, with Ti-Mn and Ti-5Al-5Cu being the lowest and the highest, respectively. Although the graph for the yield strengths corresponding to the metals shown in Fig. 2 is not

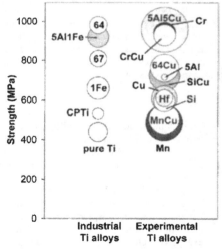

Figure 2. Tensile strengths of alloys tested. (The radius of each circle for each alloy indicates one standard deviation for each strength).

200

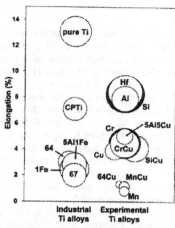

Figure 3 Elongation of metals tested

shown, they generally ranged from approximately 10% to 20% of the tensile strengths.

The percent elongation of the corresponding metals shown in Fig. 2 is given in Fig. 3. The elongation of all the cast industrial titanium alloys was in the range of approximately 1.5 to 4.0%, whereas the elongation of the cast experimental alloys scattered over a large range from about 0.5 to 10%. The Ti-Hf, Ti-Al and Ti-Si alloys had particularly high values. Adding Cu to the alloys resulted in a reduction in the ductility. Thus far, the only titanium alloys actually used in clinical dental practice are Ti-6Al-4V and Ti-6Al-7Nb. The results in Fig. 2 suggest that from the standpoint of the strength needed for high stress-bearing prostheses, industrial alloys (Ti-5Al-1Fe) and certain experimental alloys, such as ternary Ti-Al-Cu and Ti-Cr-Cu alloys, are worth testing in clinical trials.

Table 1. Chemical compositions and phases in cast industrial and experimental titanium alloys tested

Metals	Code	Chemical composition (mass %)	Alloy phase	Manufacturer
Pure Ti (sponge Ti)* (TST-1)	pure Ti	0.035Fe; 0.030 O; 0.0026N; <0.009C;	α	Toho Titanium. Co. Ltd., Japan
Industrial Ti alloys				
CP Ti (ASTM grade 2)	CP	<0.08Fe; <0.140 O; <0.006N, <0.009 C;	α	Titanium Corp., USA
Ti-6Al-4V (ASTM grade 5)	64	<6.10Al; 3.97V, <0.180Fe; <0.180 O; <0.01N; 0.010C;	α+β	Titanium Corp.. USA
Ti-6Al-7Nb (T Alloy Tough)	67	6Al; 7Nb; other 0.5;	α+β	GC Corp., Japan
Timetal 21SRx	MoNb	14Mo; 2.2Nb; 0.15Si	β	Timet, USA
Ti-13Zr-13Nb	13-13	13Zr; 13Nb	α+β	Smith and Nephew Richards, Inc., USA
Ti-15V-3Cr-3Sn-3Al	15-3	15V; 3Cr; 3Sn; 3Al	α+β	Toho Titanium. Co. Ltd., Japan
Super TIX800	1Fe	0.910Fe; 0.370 O; 0.005N, 0.004 C	α+β	Nippon Steel, Japan
Super TIX51AF	5Al1Fe	5.04Al; 1.04Fe; 0.174 O; 0.002N; 0.003 C	α-β	Nippon Steel, Japan
Experimental Ti alloys				
Ti-Cu	Cu	1, 2, 3, 4, 5, 7, 10Cu	α+Ti₂Cu	
Ti-64-Cu	64Cu	Ti-6Al-4V-1, 4, 10Cu	α+β	
Ti-Cr	Cr	7, 13, 19Cr	7Cr:α+β, >13Cr:β,	
Ti-Cr-Cu	CrCu	7Cr-3Cu, 13Cr-3, 5, 7Cu, 19Cr-3Cu	7Cr:α+β, >13Cr:β,	
Ti-Hf	Hf	5,10, 20, 30, 40Hf	α	
Ti-Si	Si	0.2, 0.5, 0.9, 1.5, 2.0Si	α	
Ti-Si-Cu	SiCu	0.5Si-2, 5Cu, 1.5Si-2, 5Cu	α+Ti₂Cu	
Ti-Al	Al	5Al	α	
Ti-Al-Cu	AlCu	5Al-1, 3, 5Cu	α+Ti₂Cu	
Ti-Mn	Mn	5, 10, 15, 18, 20, 26Mn	β	
Ti-Mn-Cu	MnCu	1.3Mn-2Cu, 3.3Mn-5Cu, 10, 18, 26Mn-2Cu	β	
Commercial dental alloys				
Ney-Oro B-2	Au	74Au; 11.5Ag; 9.5Cu; 4Pd; 1Zn		Ney Dental Inc., USA
Vitallium	CoCr	60Co; 31.5Cr; 6 0Mo		Dentsply Austenal. USA
Talladium V	NiCr	60-76Ni; 12-21Cr, 4-6Ti		Talladium Inc.,USA

* Ingots were made by melting sponge Ti in an arc-melting furnace

Castability

The majority of dental appliances are custom-made for patients' needs using lost-wax investment casting. It is important to make a sound, defect-free casting by filling the mold cavity as completely as possible with the molten metal. This empirical measure, which is often referred to as "castability" [10] or "fluidity" [11], can be determined by using various mold shapes [12]. We have successfully tested the ease of filling a wedge-shaped mold. We used wedge angles of 15° and 30° [12]. Kokubo [13] reported that the average angle of the edge of a cast dental crown was $27.6 \pm 8.9°$. The extent to which the metal fills the mold (mold filling) is determined from the gap (μm) between the tip of the casting and the theoretical acute tip of the triangle seen in a photograph of the cross section of each casting. Fig. 4 summarizes the mold filling for the two wedge angles for some of the metals listed in Table 1. There were no significant differences (p>0.05) for all the titanium tested in the 30° wedge angle. However, the mold filling of the 1% and 4% Cu alloys was significantly different (p<0.05) in the 15° wedge angle.

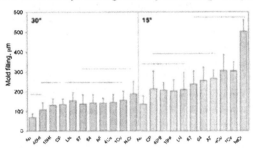

Figure 4   Mold filling of alloys tested. The horizontal bars indicate no statistical differences.

The way the mold is filled is reportedly governed by the aspects of filling: "flowability", which is limited by heat transfer, and "fillability", which is limited by surface tension [11]. Thus, in thin-section molds, both heat transfer and surface tension combine to limit the filling of the mold. The fluidity is influenced by the metal variables, mold, and mold-metal variables and equipment variables. The latter two variables in our study are considered to be similar for all the titanium metals tested. The metal variables include surface tension, kinematic viscosity, fusion temperature vs. mold temperature, and liquidus and solidus temperatures. Among these, one factor differentiates the Ti-Cu alloys from the other titanium: the temperature difference between the liquidus and solidus temperatures in these alloys [6]. The pasty-freezing [11] created by solidification over a temperature range increased the viscosity, resulting in poorer mold filling compared to other titanium [12]. The future challenge is to find other alloying elements that may enhance the fluidity without sacrificing some of the benefits of Cu as an alloying element.

The mold filling of the precious alloy (Au-Ag-Cu-Pd) tested was the best. Our earlier speculation [12] about the differences in the surface tension and kinematic viscosity indicated favorable mold filling for the gold alloy. The inferior mold filing of the Ni-Cr alloy is difficult to explain. There is a large difference (3 times) in the centrifugal force used in casting titanium and Ni-Cr alloy [14] which affected the mold filling of the Ni-Cr alloy to some extent, but the centrifugal force used to cast the gold alloy was not as high as for the titanium casting.

Grindability

Several types of abrasive wheels and burs are used in dentistry to grind, cut and polish castings. These processes are necessary for completing dental appliances. Their efficiency is directly related to the cost of the prostheses; thus, how easily and quickly cast metal can be finished is important from the viewpoint of cost effectiveness. The grindability and machinability of titanium is generally considered to be poor compared to many common metals and alloys due to several properties of titanium including high chemical reactivity, relatively lower thermal conductivity, high strength at high temperature, and low modulus of elasticity [15]. However, we

202

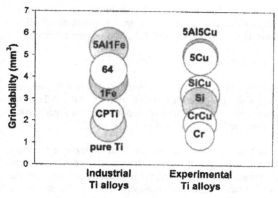

Figure 5 Results of grindability testing

found that a much better grindability than is typical for pure titanium can be attained by carefully selecting the alloying elements. To obtain comparable numbers, we exam-ined grindability [the volume loss (mm³ after 1 min grinding] by grinding titanium specimens with an SiC wheel at the rotational speeds of 500-1,250 m/min (circumferential speed) at 100 gf [9]. Fig. 5 compares the grindability at 1,250 m/min of several cast industrial and experimental titanium alloys. In both alloy groups, the values ranged from approximately 1 to 5.5 mm³. Some prominent differences in grindability were found between types of metals. Of the industrial titanium alloys, pure titanium and CP Ti have low grindability, whereas other industrial titanium alloys tested had better grindability; the value for the Ti-5Al-1Fe was the best. The Ti-5Al+Cu (1-5%) alloys had the best results of the experimental alloys with 5% Cu being the best. The trend of improving grindabilty by adding Cu is seen in Fig. 5. There were some trends in the grindability in terms of the microstructures and existing phases of the alloys.

To find the mechanism to improve grindability, Chan et al. [16] analyzed the grindability of titanium alloys by considering the fracture behavior of alloys in response to the stress field of a grinding wheel, which was approximated as a cylindrical disk with a flat region acting on a flat substrate. The initiation and propagation of microcracks in the substrate was examined on the basis of one of two fracture criteria: (1) a critical stress criterion for the initiation of cleavage cracks, and (2) a critical stress intensity factor (K_{IC}) criterion for the initiation and propagation of shear cracks. By relating K_{IC} to tensile properties through a J-integral analysis, Chan et al. [16] derived a relation between grindability (G_R), grinding speed (s), and tensile properties as follows:

$$G_R = \frac{\alpha' s}{t\pi}\left[\frac{L}{1.18\sigma_{UTS}\sqrt{\pi a}\left(E/\sigma_{ys}\right)\left[\varepsilon_y\left(1+J_z/J_e\right)+h(v,n)\varepsilon_p\right]}\right]^2 \qquad (1)$$

where L is the applied load on the grinding wheel and t is the wheel thickness. E is Young's modulus, σ_{ys} is the yield stress, σ_{UTS} is the ultimate tensile strength, ε_y is the yield strain, and ε_p is the plastic strain at fracture or tensile ductility. The parameter n is the inverse of the strain hardening exponent, α' is an empirical constant with a value of 100, and h is a function of the n and the Poisson's ratio, v. The J_z/J_e ratio is a function of applied stress to the yield stress ratio and n; its expression is given in Chan et al. [16].

The grindability model, Eq. (1), indicates that G_R increases with increasing grinding speed but increases with decreasing σ_{UTS} and ε_p values. Grindability was computed as a function of grinding speed and microstructure for cast titanium alloys containing an α, $\alpha + \beta$, or β microstructure with or without intermetallic eutectoid structures. Model predictions indicated that the grindability of titanium alloys increases with grinding speed but increases with decreasing fracture toughness or tensile ductility. The theoretical results agree with experimental data. The comparison with experimental data also revealed that alloying additions leading to the formation of intermetallics with reduced ductility enhances the grindability by reducing fracture toughness, tensile ductility, and the resistance to crack initiation and propagation. Apparently,

203

the grindability is favorable when the metals have low ductility and a multiphase structure ($\alpha+\beta$) and/or eutectoid structure [17]. Thus, highly ductile metals such as pure titanium and β alloys (for example, Ti-15V-3Cr-3Sn-3Al and the Ti-Cr alloys) [18] do not have good grindability.

Wear Behavior

As in the case of grindability/machinability, titanium is also noted for its poor tribological characteristics [15]. In his clinical study, Kabe [19] reported severe wear of cast CP Ti teeth. The wear resistance of dental restorative materials is of utmost importance since good wear resistance contributes to stable, long-term restored occlusion [19].

In most cases in the titanium industry, surface coating processes are used to improve the wear behavior of titanium [15]. In dentistry, however, such surface treatments are not practical because of the additional steps required to make a prosthesis and the reduced effectiveness of the surface-treated layers during clinical use. Our investigation indicated that wear is similar to grindability with respect to the finding that the wear resistance of titanium can be improved by alloying. The amount of wear of the cast titanium teeth was investigated using two-body wear testing equipment that simulates chewing action [20]. Testing was performed by repeatedly applying a load of 49 N (5kgf) to cast maxillary and mandibular teeth for 60 cycles/min under continuous water spray. Wear was assessed as volume loss after each set of teeth underwent 50,000 cycles. Fig. 6 summarizes the results of our wear evaluation for cast commercial titanium and experimental alloys.

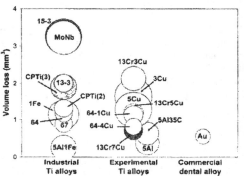

Figure 6 Wear results for metals tested

The ductile commercial alloys such as 15-3 and Timetal (MoNb) with BCC structure displayed poor wear resistance, followed by CP Ti [20]. Other metals with two-phase $\alpha+\beta$ microstructure had better wear resistance. Alloying Cu was effective at improving the wear resistance of the experimental alloys; Cu added to Ti-6Al-4V further improved its wear resistance. These results indicate that the less ductile $\alpha+\beta$ alloys and the alloys with a eutectoid microstructure had improved wear. In an earlier examination of cross sections showing the microstructures near the surfaces of titanium teeth specimens, a massive plastic deformation occurred near the surfaces where the wear processes took place [20]. Thus, we believe that any microstructural modifications that restrict the plastic deformation can improve the wear characteristics.

Similar to the grindability analysis, the wear behavior was analyzed using the elastic-plastic fracture of individual alloys in response to the relevant contact stress field [21]. Using the contact stresses as the process driving force, wear was computed as the wear rate (volume loss) as a function of hardness and tensile ductility for cast titanium alloys with an α, $\alpha + \beta$, or β microstructure with or without intermetallic eutectoid structures. The total wear after N cycles has been obtained as [21]:

$$ W = c_o \varepsilon_p N^{1+2\beta} \left[\frac{\sigma}{H} \right]^{3/2} $$

(2)

where σ is the applied pressure on the wear pad, H is the hardness, ε_p is the tensile ductility of the substrate, and c_o and β are empirical constants evaluated from experimental data [21]. Model predictions indicated that wear of titanium alloys increases with increasing hardness but increases with decreasing fracture toughness or tensile ductility; both are in agreement with experimental data of Ti alloys. The wear data of α, α+β , and β Ti alloys with a eutectoid are comparable and all exhibit higher wear resistance than the single-phase β Ti alloys. In α+β titanium and precipitation-hardened β Ti alloys, the inclusion of a two-phase structure, eutectoid structures as well as the precipitated phases, enhances wear resistance by reducing the tensile ductility and increasing hardness. Similarly, the acicular α' martensite microstructure (possibly in some Ti-Cr alloy series) [18], which exhibits a lower tensile ductility and a higher hardness, is more beneficial for wear resistance compared to the equiaxed β microstructure in single-phase β Ti alloys.

Corrosion Behavior and Biocompatibility

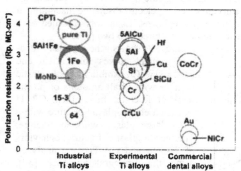

Figure 7 Corrosion behavior of tested metals [27]

It is well documented that pure titanium has extremely low toxicity and is well tolerated by both bone and tissue [23]. One important issue to be addressed is the effect of alloying elements on the corrosion characteristics and cytotoxicity. There is a concern that "diluting" titanium with alloying elements might jeopardize its outstanding corrosion behavior and excellent cytotoxicity. We examined many of our experimental alloys using a dynamic polarization method (in artificial saliva), in which open circuit potential (OCP) value, polarization or corrosion resistance, corrosion current density and passivation current density were evaluated [9, 22, 24, 25, 26]. We consistently found that the electrochemical behavior of all of the titanium alloys tested was similar to that of pure titanium within the internal oxidation potential. Some quantitative differences in the polarization resistance for the commercial titanium alloys and experimental alloys are summarized in Fig. 7. Some superior polarization resistance results were similar to those for pure titanium and CP Ti; overall the anodic polarization showed no unusual activity in any tested titanium alloys within the normal intraoral potential. The polarization resistance of Ti-6Al-4V had the lowest value (Fig. 7) of all the titanium tested because passivation films formed on the surfaces of the titanium alloys tested (similar to the films on pure titanium), depending on the composition. The titanium in the alloys is thermodynamically oxidized preferentially compared to most of the alloying elements [27].

One notable finding is the low polarization resistance of the precious commercial alloys tested and the Ni-Cr alloy. Apparently in this testing environment, which is commonly used for the corrosion evaluation of titanium, the surfaces of the gold-based alloy tarnished considerably, whereas the surfaces of the Ni-Cr alloy were covered by corrosion products. We believe these phenomena were caused by the dissolution of the alloying elements such as Ag, Cu and Ni, which is indicated by an early breakdown in their anodic polarization curves. The outstanding corrosion resistance of titanium alloys almost assures good biocompatibility. Nevertheless, the cytotoxicity of selected industrial and experimental titanium alloys was tested [28]. The test method is similar to that used earlier by Watanabe et al. [29] for other titanium alloys. Polished

205

specimens were placed directly with Balb/c 3T3 fibroblasts for 72 hrs. The cytotoxicity was evaluated by the MTT method, using CP Ti, Ti-6Al-4V and Teflon as controls, according to ISO 10993-5. The alloys tested included Ti-1Fe, Ti-5Al-1Fe, Ti-Mo-Nb, Ti-13Cr-5Cu, and Ti-1.5Si-5Cu in addition to CP Ti and Ti-6Al-4V. The cytotoxicity expressed as values normalized to the Teflon control ranged from 82% to 96% [27]. We believe that the experimental alloys tested pose no greater cytotoxic risks than do many other commonly used alloys including CP Ti.

Fracture Toughness

The fracture toughness of alloys is influenced by alloy composition and microstructure. Since high fracture toughness and tensile ductility lead to poor grindability, alloy composition and microstructure must be selected judiciously so that titanium castings exhibit balanced properties of good grindability, high strength, fracture toughness, wear resistance and tensile ductility.

Fracture toughness tests were performed at 25°C by three-point bending in a servo-hydraulic testing machine (MTS System, Minneapolis, MN) at a loading rate of 60 N/s. The rectangular bars were about 35 mm long, 7 mm wide, and 5 mm thick. Using electro-discharge machining, a 3.5 mm-long notch was made at the center of the test specimens. Load was recorded as a function of the load-line displacement until the test specimen fractured into two pieces.

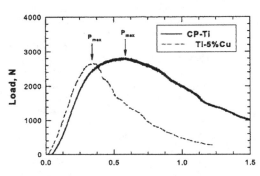

Figure 8 shows the typical load-displacement curves for Ti-5Cu and CP Ti. Ti-5Cu exhibited mostly elastic fracture with small amounts of plastic work at fracture. In contrast, CP Ti exhibited elastic-plastic fracture with a fairly large plastic work at fracture. The elastic-plastic fracture behaviors of Ti-5Cu and CP Ti were analyzed using the J-integral approach [30]. Following ASTM procedures [31,32], the maximum load, P_{max}, in the load-displacement curve was utilized to compute the stress intensity factor, K_q, at the onset of fracture. The elastic component, J_e, and the plastic component, J_p, of the J-integral were computed using the load-displacement data. Once J_e and J_p were computed, the critical stress intensity factor, K_{JC}, was calculated on the basis of the J-integral at fracture [30]. The K_q and K_{JC} values for Ti-5Cu are 49.96 ± 0.46 $MPa(m)^{1/2}$ and 63.22 ± $MPa(m)^{1/2}$, respectively, compared to 42.0 ± 2.92 $MPa(m)^{1/2}$ and 63.22 ± $MPa(m)^{1/2}$ for CP Ti. The lower fracture resistance in Ti-5Cu is the result of Ti_2Cu eutectoid particles that serve as crack initiation sites and promote cleavage.

Figure 8. Load vs. load-line displacement curves for 3-point bend fracture toughness testing of Ti-5Cu and CP Ti at ambient temperature.

Fatigue Life

The surfaces of small dental titanium devices are typically polished and smooth. The fatigue life is likely to be spent mostly in crack initiation; once initiated, the fatigue crack does not need to propagate far to reach a critical size by virtue of the small device dimensions.

The fatigue life of the titanium alloys was studied by three-point bending of rectangular bar specimens (35 mm long, 7 mm wide, 6 mm thick) that were machined from rectangular cast bars by electron-discharge machining and subsequently mechanically polished down to 400 grit SiC papers. The fatigue tests were performed in a MTS testing machine at 10 Hz under load-controlled conditions at a maximum stress of 400 MPa on the outer surface and an R ratio of 0.1, where R is the ratio of the minimum stress to the maximum stress in the fatigue cycle. The fatigue test was conducted until the specimen failed into two pieces or the number of cycles exceeded about 1.3 million.

The number of fatigue cycles to failure was taken to be the fatigue life. The Ti-5Cu specimens were tested at a maximum stress of 400 MPa and a stress ratio, R, of 0.1. The fatigue life ranged from 20 kcycles to more than 1340 kcycles (416.1 ± 626.4 kcycles). The CP Ti specimens were examined under the same maximum stress and R ratio. The fatigue life ranged from 56 kcycles to more than 1350 kcycles (504.5 ± 735.0 kcycles). Inspection of the fracture surfaces by optical microscopy indicated that the fatigue cracks initiated from the corners of the notch tip of the three-point bend specimens for both Ti-5Cu and CP Ti. The possible causes of the large variations in fatigue life are still being investigated.

Summary

We have examined some characteristics of inexpensive titanium alloys in an effort to develop alloys for use in dental prostheses. The results thus far are encouraging: the alloys are easy to cast, and they show good corrosion resistance and biocompatibility. Future studies must be done to improve their casting efficiency and accuracy, their bonding ability with porcelain, and their machinability for CAD/CAM applications. The development of these alloys looks promising with further refinements to alloys and casting systems.

Acknowledgments

The authors appreciate the efforts of the dedicated research collaborators who contributed to the referred articles. They also appreciate the editorial assistance by Jeanne Santa Cruz. This study was partially funded by National Institutes of Health/National Institute of Dental Research grant DE11787.

References

1.   T. Okabe and H. Hero,"The Use of Titanium in Dentistry," *Cells and Materials,* 5 (1995), 211-230.
2.   T. Okabe et al., "The Present Status of Dental Titanium Casting," *Journal of Metals*, 50 (1998), 24-29.
3.   C. Leyens and M. Peters, *Titanium and Titanium Alloys: Fundamentals and Applications.* (Weinheim: Wiley VCH GmbH & Co. KGaA, 2003).
4.   M. Niinomi et al. "New β-type Titanium Alloys with High Biocompatibility," *Non-aerospace Applications of Titanium,* ed. F.H. Froes, P.G. Allen, M. Niinomi (Warrendale, PA: The Minerals, Metals & Materials Society, 1998).
5.   Y. Okazaki, et al., "Corrosion Resistance, Mechanical Properties, Corrosion Fatigue Strength and Cytocompatibility of New Ti Alloys Without Al and V," *Biomaterials,* 19 (1997), 1197-1215.
6.   J.L. Murray, *"Phase Diagrams of Binary Titanium Alloys"* (Metals Park, OH: ASM International, 1987).
7.   M. Morinaga, N. Yukawa, and H. Adachi, "Electronic Structure and Phase Stability of Titanium Alloys," *J Iron and Steel Institute Japan*, 72 (1986), 555-562.
8.   R.I. Jaffee. "The Physical Metallurgy of Titanium Alloys," *Progress in Metal Physics. Vol. 7,* ed. B. Chalmers (London:Pergamon Press, 1958), 65-163.

9. M. Koike, et al., "Evaluation of Cast Ti-Fe-O-N Alloys for Dental Applications," *Mater Sci Eng C*, 25 (2005), 349-356.

10. M.C. Flemings, *Solidification Processing. Vol. 163* (New York: McGraw-Hill, 1974), 219-224.

11. J. Campbell, *Castings* (Oxford: Butterworth Heinemann, 1991).

12. H. Shimizu et al., "Mold Filling of Titanium Alloys in Two Different Wedge-shaped Molds," *Biomaterials*, 23 (2002), 2275-2281.

13. S. Kokubo, "Studies of Marginal Castability of Cast Crown," *J Med Dent Sci*, 59 (1992), 33-47.

14. I. Watanabe et al., "Effect of Pressure Difference on the Quality of Titanium Casting," *J Dental Research*, 76 (1997), 773-779.

15. R. Boyer, G. Welsch, and E.W. Collings, *Materials Properties Handbook: Titanium Alloys* (Metals Park, OH: American Society for Metals, 1994).

16. K.S. Chan, M. Koike, and T. Okabe, "Grindability of Ti Alloys," *Metall Mater Trans A*, 87A (2006), 1323-1331.

17. T. Okabe et al. "Improving Grindability and Wear Resistance of Titanium Alloys," *Cost-affordable Titanium Symposium Dedicated to Professor Harvey Flower*, ed. F.H. Froes, M. Ashraf Imam, D. Fray (Warrendale, PA:The Minerals, Metals & Materials Society, 2004), 177-181.

18. M. Koike et al., "Evaluation of Ti-Cr-Cu Alloys for Dental Applications," *J Materials Engineering and Performance*, 14 (2005), 778-783.

19. S. Kabe, "Studies on Attrition of CP Titanium as Metal Teeth," *Tsurumi University Dental Journal*, 24 (1998), 69-79.

20. C. Ohkubo et al., "In Vitro Wear Assessment of Titanium Alloy Teeth," *J Prosthodontics*, 11 (2002), 263-269.

21. K.S. Chan, M. Koike, and T. Okabe, "Modeling of Wear of Cast Ti Alloys," *Acta Biomaterialia* (accepted).

22. M. Koike, et al., "Evaluation of Cast Ti-Fe-O-N alloys for Dental Applications", *Journal of the Japan Institute of Light Metals*, 55 (2005), 682-686 (in Japanese).

23. D.I. Bardos. "Titanium and Titanium Alloys", *Concise Encyclopedia of Medical and Dental Materials*, ed. D. Williams (Oxford: Pergamon Press, 1990), 360-365.

24 M. Koike et al., "Corrosion Behavior of Cast Ti-6Al-4V Alloyed with Cu," *J Biomaterials Materials Research*, 73 (2004), 368-74.

25. Z. Cai et al., "Electrochemical Characterization of Cast Ti-Hf Binary Alloys," *Acta Biomaterialia*, 1 (2005), 353-356.

26. M. Koike et al., "Corrosion Behavior of Cast Titanium with Reduced Surface Reaction Layer Made by a Face-coating Method," *Biomaterials*, 24 (2003), 4541-4549.

27. Y. Takada et al., "Microstructure and Corrosion Behavior of Binary Titanium Alloys with Beta-stabilizing Elements," *Dental Materials Journal*, 20 (2001), 34-52.

28. M. Koike et al., "Cytotoxity of Novel Titanium Alloys," *J Dental Research*, 85 (Special Issue B) (2006), No. 2555.

29. I. Watanabe, "Cytotoxicity of Commercial and Novel Binary Titanium Alloys With and Without a Surface-reaction Layer," *J Oral Rehabilitation*, 31 (2004), 185-189.

30. J.R. Rice, "A Path Independent Integral and the Approximate Analysis of Strain Concentration by Notches and Cracks," *J Appl Mech*, 35 (1968), 379-386.

31. ASTM Standard, E 1739-96, Standard Test Method for J Integral Characterization of Fracture Toughness, Annual Book of ASTM Standards, Section 3, Vol. 03.01, 957-980.

32. ASTM Standard, E 1820-96, Standard Test Method for Measurement of Fracture Toughness, Annual Book of ASTM Standards, Section 3, Vol. 03.01, 981-1013.

MECHANICAL PROPERTIES OF α + β TYPE TITANIUM ALLOYS FABRICATED BY METAL INJECTION MOLDING WITH TARGETTING BIOMEDICAL APPLICATIONS

Mitsuo Niinomi[1], Toshikazu Akahori[1], Masaaki Nakai[1], Kazuma Ohnaka[2], Yoshinori Itoh[3], Kenji Sato[3], Tomoya Ozawa[4]

[1]Department of Biomaterials Science, Institute for Materials Research, Tohoku University; 2-1-1, Katahira, Aoba-ku, Sendai, 980-8577 Japan
[2]Graduate Student of Toyohashi University of Technology; 1-1, Hibarigaoka, Tempaku-cho, Toyohashi, 441-8580 Japan
[3]Hamamatsu Industrial Research Institute of Shizuoka Prefecture; 1-3-3, Shinmiyakoda, Hamamatsu, 431-2103 Japan
[4]Teibow Co., Ltd.1-2-1, Mukojuku, Hamamatsu-shi, Shizuoka-ken, 430-0851, Japan

Keywords: Metal injection molding, Ti-6Al-4V, Ti-6Al-4V-3Mo, Biomedical application, Mechanical properties

Abstract

Metal injection molding (MIM) is a cost-saving process for fabricating products that have complicated shapes with a very high accuracy of size. It is expected that MIM will be employed as a new process for fabricating orthopedic and dental products with complicated shapes.

The mechanical properties of Ti-6Al-4V, particularly its tensile and fatigue properties, fabricated by MIM (Ti64) were investigated to determine their suitability for medical applications. In order to improve the reliability of Ti64 for biomedical applications, the effects of heat treatments, and the addition of an element (Mo) on its mechanical properties were also investigated.

The 0.2% proof stress, tensile strength and elongation of Ti64 are approximately 740 MPa, 850 MPa, and 12%, respectively. The volume fraction of the pores is around 1.1%. The tensile strength of Ti64 subjected to heat treatment at 1323 K for 3.6 ks followed by air cooling and hot rolling increases by around 24%. However, it is difficult to improve the strength-elongation balance because the elongation decreases after heat treatment and hot rolling. The strength-elongation balance of Ti64 is improved by hot isostatic pressing (HIP): the tensile strength and elongation are around 960 MPa and 13%, respectively. The tensile strength of Mo added Ti64 that contains Mo (Ti-6Al-4V-3Mo) increases by around 15% in comparison with that of Ti64. The elongation of both the alloys is almost the same. The size of the pores of Ti64 is reduced drastically by HIP. The fatigue strengths of Ti64 subjected to HIP is the highest in the low- and high-cycle fatigue-life regions. Its fatigue strength is comparable to that of wrought Ti-6Al-4V.

Introduction

Titanium (Ti) and its alloys have attracted considerable attention for medical applications, particularly implants such as artificial hip joints, bone plates, and dental implants. Among the titanium alloys developed to date, Ti-6Al-4V ELI is extensively used as an implant material. The machinability of Ti alloys is very low. Since the implants have complicated shapes, their machining cost is very high and therefore they are very expensive. Near net shape forming

209

processes such as precision casting and powder metallurgy are effective in solving the problem of cost. Precision casting process is a promising process for developing implants with complicated shapes [1] [2]. However, the following problems tend to occur in precision casting: element segregation, significantly large casting defects such as shrinkages and microstructure coarsening, and severe reactions between the melt and the mold material. Metal injection molding (MIM), in which slurry comprising metallic and resin powder is injected into a metallic die, is effective in avoiding these defects. The density of products fabricated by MIM is generally above 95%, although they contain very small pores [3] [4]. It is expected that MIM will be employed as a new forming process for fabricating orthopedic and dental products [5] [6].

Plate specimens of Ti-6Al-4V for tensile and fatigue tests were fabricated by MIM. In order to improve their mechanical properties, they were subjected to various post heat treatments followed by the addition of Mo. The tensile and fatigue properties of each plate specimen were investigated. Then the applicability of MIM process for fabricating metallic implants was discussed.

Experimental Procedures

The following powders were used in this study: gas-atomized pure Ti powders with an average diameter of 45 μm and a purity of 99.7%, Al-V alloy powders with an average diameter of 45 μm, which were fabricated by crashing Al-40% V mother alloy produced by the thermit process using jet mill crashing; and Mo powders with an average diameter of 1.6 μm and a purity of 99.90%. Those powders were mixed to obtain the chemical compositions of Ti-6Al-4V and Ti-6Al-4V-3Mo. The weight of the total powder of each alloy composition was 2610 g. Each alloy powder batch was pre-mixed for 3.6 ks in air using a ball mill and then kneaded with binder at 443 K for 8.1 ks using a pressure-type kneading machine. The binder contained polypropylene, acrylic resin, paraffin wax and stearic acid in a mixture ratio of 30: 40: 29: 1. The resulting compounds were injection molded into compacts for tensile test and fatigue tests shown in Fig.1. Injection molding was carried out at 443 K under a pressure of 65 MPa. The compacts were subjected to the extraction debinding at 343 K for 21.6 ks using n-hexan in order to partially remove the binder. Subsequently, thermal debinding was carried out on the compacts in argon gas with a purity of 99.99 % under a pressure of 800 Pa. Sintering was performed at 1473 K for 28.8 ks in a vacuum (10^{-2} Pa) followed by furnace cooling. The sintered compacts of Ti-6Al-4V and Ti-6Al-4V-3Mo were termed Ti64 and Ti643, respectively. Some Ti64 specimens were subjected to solution treatment at 1313 K for 3.6 ks in an argon atmosphere followed by air cooling, as shown in Fig.2 (a): hot isostatic pressing (HIP) in which they were holed at 1123 K for 7.2 ks followed by furnace cooling, as shown in Fig.2 (b); or hot rolling at 823 K in air a reduction ratio of around 30 % as shown in Fig.2 (c). The Ti64 specimens subjected to solution treatment, HIP, and hot rolling were termed $Ti64_{HA}$, $Ti64_{HIP}$, and $Ti64_{HR}$, respectively.

The tensile test was conducted on each sintered specimen, as shown in Fig. 1 (a), by using an Instron-type machine at a crosshead speed of 8.33×10^{-6} in air at room temperature. The fatigue test was also conducted on each sintered specimen, as shown in Fig.1 (b), by using an electro-servo-hydraulic fatigue-testing machine at a frequency of 10 Hz and a stress ration of R = 0.1 under sine wave stress conditions in air at room temperature.

Microstructural observations of each specimen were conducted using an optical microscope and fracture surface observations were conducted using a scanning electron microscope (SEM). X-ray diffraction analysis was performed at a voltage of 40 kV and 30 mA using Cu-K$_\alpha$ radiation.

210

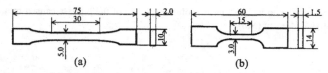

Figure 1. Geometries of (a) tensile and (b) fatigue test specimens in mm.

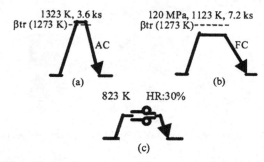

Figure 2. Schematic drawings of heat and thermomechanical treatments for Ti64.

Results and Discussion

Microstructure

The microstructures of Ti64, Ti64$_{HA}$, Ti64$_{HIP}$, Ti64$_{HR}$, and Ti643 are shown in Fig.3. The pores observed in the microstructures of Ti64, Ti64$_{HA}$, and Ti643 are relatively larger than those observed in the microstructures of Ti64$_{HIP}$ and Ti64$_{HR}$.

The microstructure of Ti64 (Fig.3 (a)) shows colonies, where coarse lamellar like acicular α phases are formed, in coarse prior β grains with an average diameter of 104 μm. The microstructure of Ti64$_{HA}$ (Fig.3 (b)) shows very fine acicular α phases in prior β grains: however, the colonies cannot be observed in these β grains. The microstructure of Ti-6Al-4V is converted into a single β phase when it is heated over the β transus temperature. Therefore, it is considered that in the case of Ti64$_{HA}$, the acicular α phases existed before heating over the transus temperature; they began to disappear when Ti64$_{HA}$ was maintained at a temperature above the β transus temperature, and very fine acicular α phases newly precipitated during air cooling. As in the case of Ti64, the microstructures Ti64$_{HIP}$ (Fig.3 (c)) and Ti64$_{HR}$ (Fig.3 (d)) show colonies, where coarse lamellar- like acicular α phases are formed, in coarse prior β grains. However, the average diameter of the prior β grains of Ti64$_{HIP}$ is 90.7 μm, which is slightly less than that of Ti64 while the prior β grains of Ti64$_{HR}$ are elongated along the rolling direction.

The microstructure of Ti643 (Fig.3 (e)) also shows colonies in coarse prior β grains. However, the lamellar-like acicular phases in these grains are finer than those in the prior β grains of Ti64. Further, the average prior β grain size of Ti643 is less than that of Ti64. The microstructure of

211

Ti643 is refined due to an increase in the volume fraction of the β phase caused by the addition of the β stabilizing element (Mo).

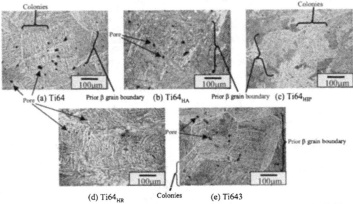

Figure 3. Optical micrographs of (a) Ti64, (b) Ti64$_{HA}$, (c) Ti64$_{HIP}$, (d) Ti64$_{H}$, and (e) Ti643.

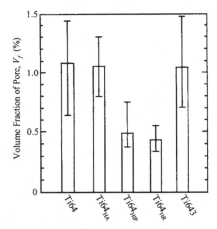

Figure 4. Volume fractions of pores in the cross sections of the tensile specimens of Ti64, Ti64$_{HA}$, Ti64$_{HIP}$, Ti64$_{HR}$, dTi643.

The volume fractions of the pores in the cross sections of the tensile specimens of Ti64, Ti64$_{HA}$, Ti64$_{HIP}$, Ti64$_{HR}$, and Ti643 are shown in Fig.4. The volume fractions of the pores observed in Ti64, Ti64$_{HA}$, and Ti643 are nearly the same. Therefore, the general heat treatment or addition of elements is not effective in reducing the volume fraction of the pores. On the other hand, the volume fractions of Ti64$_{HIP}$ and Ti64$_{HR}$ are less than those of Ti64, Ti64$_{HA}$, and Ti643.

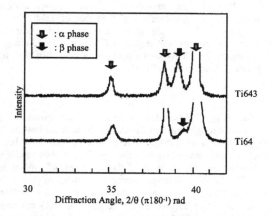

Figure 5. X-ray diffraction profiles of Ti64 and Ti643.

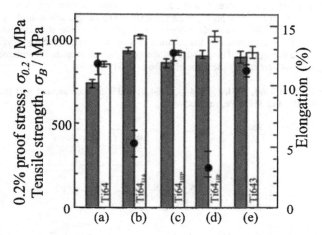

Figure 6. Tensile properties of (a) Ti64, (b) Ti64$_{HA}$, (c) Ti64$_{HIP}$, (d) Ti64$_{HR}$ and (e) Ti643.

The X-ray diffraction profiles of Ti64 and Ti643 are shown in Fig.5. The peaks of the α and β phases are observed in both profiles. The intensity of the β phase in Ti643 is stronger than that in Ti64. This implies that the volume fraction of the β phase in Ti643 is greater than that in Ti64, which was qualitatively determined by the microstructural observation.

Tensile properties

The tensile properties of Ti64, Ti64$_{HA}$, Ti64$_{HIP}$, Ti64$_{HR}$, and Ti643 are shown in Fig.6. The 0.2 % proof stress, tensile strength, and elongation of Ti64 are approximately 740 MPa, 850 MPa and

213

12%, respectively. The 0.2% proof stress and tensile strength of Ti64$_{HA}$ are greater than those of Ti64; however, its elongation is significantly less than that of Ti64. The 0.2% proof stress, tensile strength and elongation of Ti64$_{HIP}$ are significantly improved in comparison with those of Ti64. The 0.2% proof stresses and tensile strengths of Ti64$_{HR}$ and Ti643 are greater than those of Ti64. However, the elongation of Ti64$_{HR}$ is significantly less than that of Ti64, while the elongation of Ti643 is nearly equal to that of Ti64.

Fatigue properties

The S-N curves of Ti64, Ti64$_{HA}$, Ti64$_{HIP}$, Ti64$_{HR}$, and Ti643 are shown in Fig.7 along with the range of fatigue strengths of wrought and cast Ti-6Al-4V. In both the low- and high-cycle fatigue life region, the fatigue strength of Ti64$_{HA}$ is less than that of Ti64 because the ductility (elongation) of Ti64$_{HA}$ is significantly less than that of Ti64, thereby resulting in a decrease in the crack propagation resistance. However, the fatigue strength of Ti64 is in the range of the fatigue strength of cast Ti-6Al-4V. The fatigue strength of Ti64$_{HIP}$ is greater than that of Ti64 in both low- and high-fatigue life regions due to a decrease in the volume fraction and size of pores and the prior β grain size, thereby resulting in an increase in the crack initiation and propagation resistance. The fatigue strength of Ti64$_{HIP}$ is less than that of wrought Ti-6Al-4V. In the low-cycle fatigue life region, the fatigue strength of Ti64$_{HR}$ increases and exceeds that of Ti64. However, in the high-cycle fatigue life region, its fatigue strength is less than that of Ti64 due to the interaction of elongated pores that leads to stress concentration and lower ductility that results in a decrease in its crack propagation resistance. The fatigue strength of Ti643 is greater than that of Ti64 in the low-cycle fatigue life region because it increases due to the microstructural refinement caused by the addition of Mo. However, in the high-cycle fatigue life region, the fatigue strengths of Ti643 and Ti64 are nearly equal because the volume fraction and size of pores of Ti643 are nearly the same as those of Ti64.

The fatigue limits of Ti64, Ti64$_{HA}$, Ti64$_{HIP}$, Ti64$_{HR}$, and Ti643 are around 380 MPa, 280 MPa, 420 MPa, 200 MPa and 385 MPa, respectively.

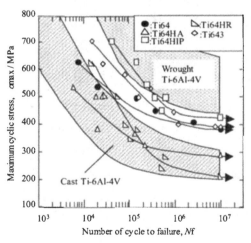

Figure 7. S-N curves of Ti64, Ti64$_{HA}$, Ti64$_{HIP}$, Ti64$_{HR}$, and Ti643.

214

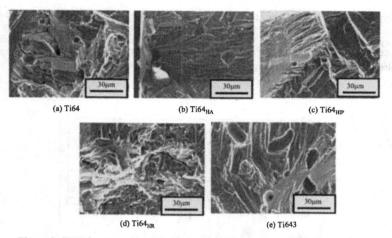

Figure 8. SEM fractographs of fatigue tested specimens of (a) Ti64, (b) Ti64$_{HA}$, (c) Ti64$_{HIP}$, (d) Ti64$_{HR}$, and (d) Ti643 in high cycle fatigue life region.

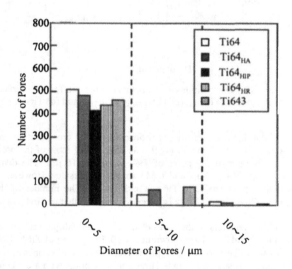

Figure 9. Distributions of diameters of pores in cross sections of fatigue test specimens of (a) Ti64, (b) Ti64$_{HA}$, (c) Ti64$_{HIP}$, (d) Ti64$_{HR}$, and (e) Ti643.

The typical fatigue fracture surfaces near the crack initiation sites of Ti64, Ti64$_{HA}$, Ti64$_{HIP}$, Ti64$_{HR}$, and Ti643 the high-cycle fatigue life region are shown in Fig.8. In both the low- and high-cycle fatigue life regions, the fatigue crack initiates from a pore located inside the specimen in Ti64, Ti64$_{HA}$, Ti64$_{HR}$, and Ti643. The fatigue crack of Ti64$_{HIP}$ does not initiate from a pore, but also from the inside of the specimen in both low- and high- cycle fatigue life regions.

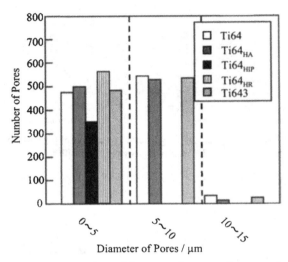

Figure 10. Distributions of diameters of pores on fracture surfaces of (a) Ti64, (b) Ti64$_{HA}$, (c) Ti64$_{HIP}$, (d) Ti64$_{HR}$, and (d) Ti643.

The distributions of the diameters of the pores observed in the cross sections of Ti64, Ti64$_{HA}$, Ti64$_{HIP}$, Ti64$_{HR}$, and Ti643 are shown in Fig.9. The diameter of most of the pores is less than 5 µm. The maximum diameter of the pores of Ti64 is around 15 µm. The distributions of the diameters of the pores of Ti64, Ti64$_{HA}$ and Ti643 are nearly the same. However, only pores with a diameter of less than 5 µm exist in Ti64$_{HIP}$ and Ti64$_{HR}$. The number of the pores with a diameter of 0-5 µm in Ti64$_{HIP}$ and Ti64$_{HA}$ is generally less than that in Ti64 and Ti643.

The distributions of the diameters of the pores observed on the fatigue fracture surfaces of Ti64, Ti64$_{HA}$, Ti64$_{HIP}$, Ti64$_{HR}$, and Ti643 are shown in Fig.10. In the case of Ti64, Ti64$_{HA}$, and Ti643, the number of pores with diameters of 5-10 µm is greater than the number of pores in the cross section of the specimens, as shown in Fig.9. However, only pores with a diameter of less than 5 µm exist in Ti64$_{HIP}$ and Ti64$_{HR}$ even on the fatigue fracture surface.

Therefore, the fatigue crack tends to initiate from a relatively larger pore. Hence, the fatigue strength of Ti64$_{HIP}$ increases and nearly equals that of wrought Ti-6Al-4V. However, the pores of Ti64$_{HR}$, are elongated and closer to each other, although their size decreases, and the interaction among them increases significantly. Therefore, the fatigue strength decreases in comparison with that of Ti64, particularly in the high-cycle fatigue life region.

Conclusions

The effects of heat treatments and the addition of an element (Mo) on the mechanical properties of Ti-6Al-4V fabricated by metal injection molding (MIM) were investigated in order to improve the reliability for biomedical applications. The following results were obtained.

(1) The 0.2 % proof stress, tensile strength and elongation of as-sintered Ti-6Al-4V (Ti64) are approximately 740 MPa, 850 MPa, and 12%, respectively.
(2) The fatigue strength and the balance of strength and ductility of Ti64 can be improved by hot isostatic pressing (HIP) such that they are comparable to those of wrought Ti-6Al-4V.
(3) The addition of Mo increases the strength of Ti64 and maintained its ductility (elongation).

References

[1] M. Niinomi et al., "Dental Precision Casting of Ti-29Nb-13Ta-4.6Zr Using Calcia Mold", *Materials Science Forum*, 475-479(2005), 2303-2308.
[2] M. Niinomi, "Mechanical Properties and Cyto-toxicity of Newly Designed Beta Type Titanium Alloys with Low Melting Points for Dental Applications", *Materials Science and Engineering C*, 25(2005), 417-425.
[3] K. Kusaka, "Tensile Behavior of Sintered Titanium by MIM Process", J. Jpn. Soc. Powder Metallurgy, 42(1995), 383-387.
[4] K. S. Scott et al., "Use of a Naphthalene-Based Binder in Injection Molding Net-Shape Titanium Components of Controlled Porosity", *Materials Transactions*, 46(2005), 1525-1531.
[5] K. Majima et al., "A Study on Sintered Titanium Alloy as a Biomaterials", *J. Jpn. Soc. for Biomaterials*, 10(1992), 3-10.
[6] Y. Itoh et al., "Fabrication of Ti-6Al-7Nb Alloys by Metal Injection Molding", Proc. Proc. PM'2006, (2006), to be published.

Mechanical Properties and Structural Superplasticity in Ultrafine Grained α-Titanium / Ti$_x$Me$_y$-Intermetallic Ti-8Fe-4Al, Ti-10Co-4Al and Ti-10Ni-4Al Alloys

[1]Georg Frommeyer

[1]Max-Planck-Institute for Iron Research, Max-Planck-Str. 1, Dusseldorf, D-40237, Germany

Keywords: superplasticity, high strain rate, grain boundary sliding, dislocation climb

Abstract

Quasi eutectoide titanium-transition metal alloys with ultrafine-grained microstructures consisting of α-titanium solid solutions, matrix grain size of about 0.5 to 0.7 micron, and a dispersion of intermetallic FeTi (B2 type of structure), Co_2Ti and Ni_2Ti (cubic complex $E9_9$ superlattice structure) particles of 0.2 to 0.3 micron in size exhibit superplasticity at high-strain rates and superior strength properties. Strain-rate-sensitivity exponents ranging between $0.4 \leq m \leq 0.5$ were achieved at high strain rates of $2 \cdot 10^{-3} \leq \dot{\varepsilon} \leq 5 \cdot 10^{-2} \text{ s}^{-1}$ and at medium temperatures from 625 to 775 °C. Maximum elongations to failure of $\varepsilon_{tot} \approx 1100$ % were recorded. TEM observations revealed relatively high dislocation densities in the α-Ti(Al) grains, whereas the intermetallic particles are nearly free of dislocations. Activation analysis showed activation energies of the order of $Q \approx 205 \pm 15$ kJ/mol. The grain size exponent was determined to be $p \approx 2.1 \pm 0.2$. The obtained results reveal clearly that superplasticity at higher strain rates in these materials is due to grain boundary sliding accommodated by dislocation climb controlled by lattice diffusion of titanium. The strong increase in strength is caused by effective particle strengthening mechanisms.

Introduction

Structural superplasticity in fine-grained metallic alloys and ceramics generally occurs at low strain rates of the order of 10^{-4} 1/s [1-4]. For economical superplastic forming operations it is required to increase the strain rate of about two orders of magnitudes. Singer and Gessinger [5] as well as Gregory, Gibeling and Nix [6] were the first who showed structural superplasticity at high strain rates in an ultrafine grained MA 6000 ODS alloy at the test temperature of 1273 K. Nieh, Henshall and Wadsworth [7] reported superplasticity at high strain rates in a SiC whisker reinforced 2124 Al alloy. Incipient melting occurred during superplastic deformation. The deformation mechanisms in this material are related to viscous flow accommodated by enhanced diffusion in the liquid/solid state. Commercial superplastic titanium alloys, such as Ti 6Al 4V, Ti 6Al 2Sn 4Zr 4Mo and SP 700Al, which show great potential applications in the aircraft and space industry, do not exhibit structural superplasticity at high strain rates at optimum deformation temperatures between 700 and 980 °C.

Two phase titanium-transition metal alloys like Ti-Ti$_x$Me$_y$ (Me:Co, Ni, Fe) with large volume fractions of the intermetallic compounds, such as Ti$_2$Co, Ti$_2$Ni and TiFe, which are finely dispersed in the h.c.p. α-titanium matrix exhibit superior superductility and strength properties. The extremely fine grained microstructure with grain and particle sizes of 0.5 and 0.2 microns promotes high-strain-rate superplasticity at relatively low deformation temperatures between 625 and 750 °C. In addition, the superior strength properties of these alloys are caused by dispersion or particle strengthening mechanisms [9-10]. The present paper describes superplasticity at high strain rates in the extremely fine grained Ti 8Fe 4Al, Ti-10Co-4Al and Ti-10Ni-4Al alloys. The results will be discussed in terms of microstructural features on the base of grain boundary sliding and dislocation creep models.

Experimental Procedure

The investigated alloys have been produced by skull-melting in an argon arc furnace using a 60 kg self-consuming electrode. The chemical composition is shown in Table 1.

Table 1. Chemical composition of the quasi eutectoid titanium alloys

	Me(Fe, Co, Ni) (wt%)	Al (wt%)	Si (wt%)	O (ppm)	N (ppm)	Ti
Ti-8Fe-4Al	8.2 ± 0.2	4.2	0.12	680	150	balance
Ti-10Co-4Al	9.8 ± 0.15	4.15	0.11	610	130	balance
Ti-10Ni-4Al	10.2 ± 0.13	4.08	0.10	630	160	balance

Table 2. Constitutional data of the binary systems investigated alloys

	C$_{eu}$(wt%) eutec.conc.	T$_{eu}$ (°C) eutec. temp.	intermet. compound	crystal structures	vol. fraction (%)
Ti-8Fe-4Al	13	630	TiFe	B2	18 ± 1
Ti-10Co-4Al	7.5	750	Ti$_2$Co	E9$_3$	23 ± 1
Ti-10Ni-4Al	5.5	800	Ti$_2$Ni	E9$_3$	22.5 ± 1

Plates from the ingot were hot rolled at 950 °C to sheets with 65 % reduction in area. A final thermomechnical treatment was carried out to refine the microstructure of the alloys with an average grain size of the α-titanium grains of about 0.5 μm and intermetallic particles with 0.3 μm in size. Microstructural investigations of the thermomechanically processed material as well as of the superplastically deformed samples were performed by optical and scanning electron microscopy (SEM) and transmission electron microscopy (TEM). Differential scanning calometry (DSC) have been carried out to determine the eutectoid transformation temperatures of the alloys. The superplastic behaviour is characterized by the strain-rate-sensitivity parameter m, defined as $m=(\partial \ln\sigma/\partial \ln\dot{\epsilon})_T$ at constant temperature and microstructure. The m-values were measured in strain-rate-change tests at different temperatures. The superplastic flow stresses of the alloys were determined in tensile tests at room and higher temperatures up to 750 °C. The Youngs' moduli are important for analyzing thermal activation processes and were measured using the resonance frequency method in the kHz range.

Results and Discussions

Microstructures

The main constitution data of the ternary phase diagrams and the lattice structures of the intermetallic compounds and their volume fractions are presented in Table 2. The ternary phase diagrams [11] of the titanium-aluminium-iron group metals show that the titanium-nickel-aluminium alloy exhibits the maximum eutectoid transformation temperature. The optimal application temperatures of the designed alloys have been established below their eutectoid transformation temperatures between 600 and 700 °C, see Table 2.

The solidified ingots reveal a typical Widmanstaetten microstructure. Hot rolling in the β region and subsequent cooling on air produces a fine lamellar troostite -pearlite with rosette morphology- type of microstructure. The eutectoid transformation mechanism have been studied in detail [12,13]. Modified thermomechanical treatment below the eutectoid transformation temperature produces a very fine equiaxed microstructure consisting of the α-Ti(Al) matrix and intermetallic TiFe, Ti_2Co, or Ti_2Ni particles, respectively. The SEM and TEM micrographs show as example in Fig. 1a and b homogeneously distributed fine Ti_2Co particles within the α-titanium-aluminium matrix. The particle diameter is about 0.2 to 0.3 μm and the α-Ti(Al) grains are between 0.5 and 0.6 μm in size.

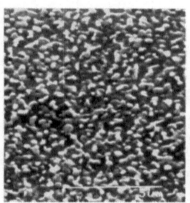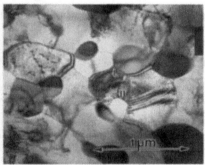

Fig. 1a-b: Representative SEM (a) and TEM (b) images of the thermomechanically processed Ti-10Co-4Al material as hot rolled in the β-phase field with 60 % reduction in area and subsequently rolled below the eutectoid transformation temperature from 730 to 600 °C.

Temperature dependent flow stress

The yield stresses of the investigated materials as a function of the test temperature are represented in Fig. 2. The thermomechanically processed Ti-10Co-4Al alloy exhibits flow stressens of the order of 1200 MPa at room temperature while maintaining sufficient room temperature ductility of 8 % elongations in tension. In the temperature range between 25 °C and 400 °C the flow stress shows a slight temperature dependence. Above 400 °C a strong decrease of the yield stress curves occurs. The Ti-10Ni-4Al alloy shows an exraordinary high flow stress at room and elevated temperature. The flow stress decreases from 1400 MPa at room temperature to 950 MPa at 500 °C. The Ti-8Fe-4Al alloy shows flow stresses of 1050 MPa at 25 °C which decrease to 500 MPa at 600 °C. It should be mentioned that the flow stress of the Ti-6Al-4V alloy is quite lower in the considered temperature region than those of the investigated α-Ti(Al) intermetallic material.

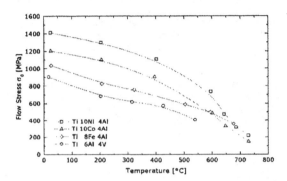

Fig. 2. Yield stress versus temperature of the quasi eutectoid titanium alloys in comparison with Ti-6Al-4V.

Superplastic Properties

For studying structural superplasticity, temperature dependent strain rate change tests have been performed in the temperature regime between 620 and 750 °C. Strain-rate-sensitivity parameters of $0.3 < m < 0.5$ have been determined in a wide strain rate and temperature range of the investigated alloys. An overview of the strain rate and temperature dependence of the m values is given in the topographic map of Fig. 3a and b for the Ti-10Co-4Al and Ti-10Ni-4Al alloys, respectively.

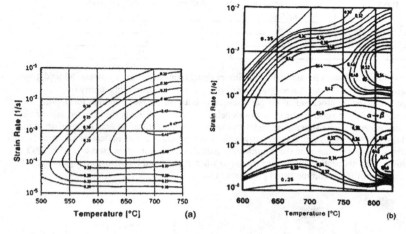

Fig. 3a-b. Topographic map of the m-values in the strain-rate-temperature regime for Ti-10Co-4Al (a) and Ti-10Ni-4Al (b).

The highest strain rate sensitivity parameter is close to the theoretical m value of 0.5. This indicates grain bounday sliding accommodated by dislocation movement involving sequential steps of glide and climb, whereas dislocation climb in the grain boundary region is the rate controlling mechanism in most fine grained materials. The m value of the so called class II solid alloys is restricted to 0.5, whereas class I solid solution alloys exceed m=0.5, typically represented by the converntional Ti-6Al-4V alloy [14]. A selected superplastically strained tensile sample of the Ti-10Co-4Al alloy is shown in Fig. 4. The sample was deformed at higher strain rates of $2 \cdot 10^{-3} \mathrm{s}^{-1}$ at 725 °C. Maximum elongation of about 1100% was achieved.

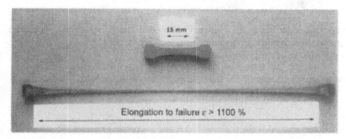

Fig. 4. Photograph of an undeformed and superplastically strained Ti-10Co-4Al tensile sample at 725 °C, initial strain rate $\dot{\varepsilon} = 2 \cdot 10^{-3} \mathrm{s}^{-1}$.

Table 3 represents the recorded elongations of the investigated alloys at high strain rates of $\dot{\varepsilon} = 3 \cdot 10^{-2} \mathrm{s}^{-1}$.

alloys	total elongations ε_{tot}(%)	deformation temperatures (°C)
Ti-8Fe-4Al	530	625
Ti-10Co-4Al	660	725
Ti-10Ni-4Al	780	750

The limiting factors for achieving maximum elongations to failure are strain induced grain growth of the α-Ti(Al) matrix grains or cavities due to the decreasing ability of accommodation processes of the particle-matrix boundaries. No cavitation formation occurred during superplastic elongations below ε_{pl}=650 % strain at the selected high strain rates because of the superior accommodation ability of the ultrafine grained microstructure. TEM micrograph in Fig. 5 shows dislocations in the matrix grains indicating slip process. No dislocations are present in the intermetallic particles. At lower strain rates of $\dot{\varepsilon}$ <10^{-3} 1/s no substantial dislocations have been observed when Friedel and Nabarro-Herring creep are the rate controlling accommodation processes [15-17].

Fig. 5. TEM image of a superplastically strained Ti-10Co-4Al sample revealing dislocations in the α-Ti(Al) grains. Deformation temperature: T=725 °C, strain rate: $\dot{\varepsilon}$=3·10^{-2}s^{-1}, elongation to failure: ε_{tot} = 660 %.

Superplastic flow is described generally by the following Dorn equation [18]:

$$\dot{\varepsilon} = A * \frac{GD_L |\vec{b}|}{kT} \left(\frac{|\vec{b}|}{d}\right)^p \left(\frac{\sigma}{G}\right)^n \tag{1}$$

where A* is a structure factor, d is the grain size, p is the grain size exponent, D_L =D_0·exp(-Q/RT) is the diffusion coefficient and n=1/m is the stress exponent. The grain size exponent describes the influence of the grain size on the strain rate. Q is the activation energy of the rate controlling diffusion process, and $\vec{b}$ is the Burgers' vector. The models for dislocation accommodation controlled by diffusion to be considered are characterized by the following parameters: n, m, p, as shown in Table 4.

Mechanism	n	m	p

Dislocation glide and climb solid solution class I behaviour	1	1	2
Dislocation glide and climb solid solution class II behaviour	2	0.5	2
Dislocation climb near grain boundary	2	0.5	1

Table 4: Stress exponent n, strain rate sensitivity exponent m and grain size exponent p for several dislocation accommodation processes.

The grain and particle size exponents have been determined for different stress levels and volume fraction of the coexisting phases. The average matrix grain size exponent is about 2.1 ± 0.2 at low and medium stress levels. From activation energy analysis based on the Arrhenius plot and using the expression $Q = -R \left(\dfrac{\partial \ln \dot{\varepsilon}/\varepsilon_0}{\partial 1/T} \right)_\sigma$ activation energies for different stress levels ($70 \text{MPa} \leq \sigma_0 \leq 400$ MPa) and temperatures between 190 and 220 kJ/mole were determined.

These data are in good agreement with the activation energy $Q = 190 \pm 10$ kJ/mole for lattice self diffusion of titanium in a α-Ti(Al) solid solution matrix, similar to those in γ-TiAl alloys [19]. From the obtained results it is deduced that high-strain-rate superplasticity in the two phase alloys of titanium-iron group metal with an ultrafine dispersion of intermetallic particles is mainly caused by grain boundary sliding and enhanced dislocation climb in the extremely fine matrix grains and explicitly in their mantle zones. The relatively low superplastic deformation temperature of these materials compared to that of the conventional Ti-6Al-4V alloy is due to a high mobility of titanium and the alloying atoms in the temperature regimes close to the eutectoid transformations in the titanium-iron group metals.

References

[1]   K.A. Padmanabhan, G.J. Davis: *Superplasticity –Mechanical and Structural Aspects, Fundamentals and Applications*, (Springer Verlag, Berlin 1980).

[2]   A. Kaibyshev: *Superplasticity of Alloys, Intermetallics and Ceramics* (Springer Verlag New York 1992).

[3]   T.G. Nieh, J. Wadsworth, O.D. Sherby: *Superplasticity in metals and ceramics* (Cambridge solid state science series, Cambridge University Press, Cambridge, UK, 1997).

[4]   F. Wakai, H. Kato: Adv. Ceram. Mat. 3 (1988), p. 71

[5]   R. Singer, G.H. Gessinger: Riso Intern. Symp. on Deform. of Polycryst. (1981), p. 365

[6]   J.K. Gregory, J.C. Gibeling, W.D. Nix: Metall. Trans. A 16A (1985), p. 777

[7]   T.G. Nieh, C.A. Henshall, J.Wadsworth: Scripta Metall. 18 (1984), p.1405

[8] R. R. Boyer, J. C. Williams, N.E Paton in: Titanium´ 99, Science and Technology, (CRISM: PROMETY, Saint Petersburg, Russia 1999), p. 1007

[9] G. Frommeyer, P. Zeitz, P. von Czarnowski in: Proc. of the Sixth World Conf. on Titanium 2 (Cannes, 1988), p. 1075

[10] G. Frommeyer, H. Hofmann, Research Report, German Science Foundation, (Bonn, Germany, 1999), p. 95

[11] J. L. Murry: Phase Diagrams of Binary Titanium Alloys, ed. by T.E. Massalski (ASM International, Metals Park, Ohio, USA,   1987).

[12] P. Zeitz, J. E. Wittig, G. Frommeyer: Scripta Met. 20 (1986), p. 757

[13] H. Hofmann and S. Strauss, Diploma Thesis, to be published 2004

[14] H. Fukuyo, H.C. Tsai, T. Oyama, O. D. Sherby: ISIJ International 31 (1991), p. 76

[15] S. Friedel: *Dislocations* (Pergamon Press, Oxford, UK 1964), p. 313.

[16] C. Herring: J. Appl. Phys. 21 (1951), p. 437

[17] F. R. N. Nabarro: Phil. Mag. A 16 (1967), p. 231

[18] O. D. Sherby and P. M. Burke: Prog. Mater. Sc.13 (1967), p.325

[19] G. Frommeyer and S. Knippscheer in: Titanium 2003 (Wiley-VCH, Weinheim, Germany, in press 2004).

DISLOCATIONS IN AMBIENT TEMPERATURE CREEP OF HCP METALS

Tetsuya Matsunaga[1, 2], Eiichi Sato[1], Kazuhiko Kuribayashi[2],

[1]ISAS/JAXA
(Institute of Space and Astronautical Science / Japan Aerospace Exploration Agency);
3-1-1 Yoshinodai; Sagamihara, Kanagawa, 229-8510, Japan
[2]The University of Tokyo;
7-3-1 Hongou; Bunkyo, Tokyo, 113-8656, Japan

Keywords: Ambient temperature creep, HCP metals, TEM

Abstract

Metals with a hexagonal closed packed (HCP) structure show creep behavior at ambient temperature. Features of this phenomenon are: (1) it appears in all and only HCP structure metals and alloys; (2) dislocations are contributing; and (3) it shows very low apparent activation energy (ca. 10 kJ/mol). Transmission electron microscope (TEM) observation was conducted on crept specimens of commercially pure Ti (CP-Ti), pure Mg, and Zn. Results showed no dislocation tangle. The dislocation arrays were aligned straightly inside the grain. The dislocation array consisted of one dislocation type. One slip system was activated in the ambient temperature creep condition. Therefore, it was considered that work hardening does not occur, and that creep deformation continued.

Introduction

Titanium alloys are beneficial materials for aerospace systems because of their high strength-to-weight ratio, high corrosion resistance, and capability of superplastic blow forming. For example, a Ti-6Al-4V alloy fuel tank was developed and used for the scientific satellite "HAYABUSA" by ISAS/JAXA in 2003. During the tank proof test, creep behavior was observed at an ambient temperature and under stresses below the 0.2% proof stress ($\sigma_{0.2}$). Although small creep strain in a fuel tank might be permissible, creep in fasteners cannot be permitted because it would cause stress relaxation, which might engender a very dangerous state.

Several studies examined ambient temperature creep in Ti-5Al-2.5Sn [1] and Ti-6Al-4V [2] in the 1960s. However, such studies ceased, except for one that addressed Ti-0.4Mn [3] in the 1990s. In the past few years, the Mills group began studying phenomena using Ti-6Al and

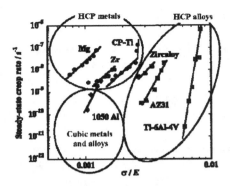

Figure 1. Double logarithmic plot of the steady-state creep rate and the modulus-compensated stress. The steady-state creep rate was obtained through fitting of the logarithmic creep eq. [6].

Ti-6AL-2Sn-4Zr-2Mo [4]. The ambient temperature creep mechanism is inferred to be related to Andrade creep [5] and deformation twinning [3]. Recently, straightly aligned dislocation arrays that are formed by solute Al atoms have been suggested [4], but those are presuppositions based only on individual experiments that are unique for tested materials.

We performed creep tests using typical HCP metals and alloys, including CP-Ti, pure Mg, Zn, Ti-6Al-4V, AZ31, Zircaloy, and some cubic metals and alloys at ambient temperature. Figure 1 shows a double logarithmic plot of the steady state creep rate and modulus compensated stress. We can categorize the data plots into three groups based on the crystallographic structure and alloying of the materials. One group has low stress and a high strain rate, which includes pure metals of HCP. Another group, that of solid solution alloys of HCP, includes materials with a high stress and a high strain rate. That of low stress, which is almost out of the figure range, is for cubic metals and alloys.

In this paper, representative HCP metals were selected as specimens: CP-Ti, pure Mg, and Zn. The HCP metals have their own c/a ratios and primary slip systems. Creep tests were performed at several temperatures to study the temperature dependency of the deformation mechanism. Finally, TEM observation was done for the three crept metals and their dislocations were characterized.

Experimental procedure

The specimens were representative HCP metals: CP-Ti, pure Mg, and Zn. Their chemical compositions, pre-treatments, c/a ratios, and primary slip systems, which are normally observed in ambient condition tensile tests, are listed in Table I. The average grain sizes were 75 μm for CP-Ti, 121 μm for Mg, and 61 μm for Zn by the intercept method.

Table I. Chemical Composition, Pre-Treatment, c/a Ratio, and Primary Slip System for the Ambient Temperature Tensile Test

Sample	Chemical Composition	Pre-Treatment	c/a	Primary Slip System
CP-Ti	Ti-0.2Fe-0.15O-0.013H-0.08C-0.03N	973 K Air Cooling	1.589	$<\bar{1}2\bar{1}0>(10\bar{1}0)$ Prismatic Slip
Mg	99.95%Mg	573 K Hot Rolling	1.624	$<\bar{1}2\bar{1}0>(0001)$ Basal Slip
Zn	99.5%Zn	Hot Rolling	1.856	$<\bar{1}2\bar{1}0>(0001)$ Basal Slip

Tensile tests were performed for deciding 0.2% proof stresses at ambient temperature with a constant crosshead speed corresponding to the strain rate of 1×10^{-2} s^{-1}. Creep tests were performed using a dead load creep frame with loads of 0.55–0.9 of the 0.2% proof stresses determined according to the tensile test results. Creep strain was measured using strain gauges with resolution of 3×10^{-6}. The test temperatures were 203–873 K.

For TEM observation, samples were prepared using twin-jet electro polishing after mechanical grinding below 50 μm thickness. The polishing solution was 6% perchloric acid, 34% 1-butoxyethanol, and 60% methanol. The polishing temperature range was 228–238 K and the respective voltages were 50 V for Mg and 10 V for Zn. For CP-Ti, ion-milling was adopted because hydride particles were deposited during electro-polishing. The operating voltage was 300 kV.

Results

The obtained 0.2% proof stresses were 220 MPa for CP-Ti, 61 MPa for Mg, and 44 MPa for Zn. Figure 2 shows examples of creep curves of the three materials at ambient temperature. All metals showed remarkable creep behavior. Figure 3 shows steady-state creep rates that were obtained through fitting of the logarithmic creep eq. (1).

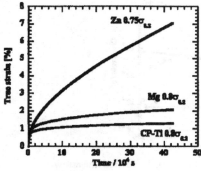

Figure 2. Creep curves of CP-Ti, Mg, and Zn.

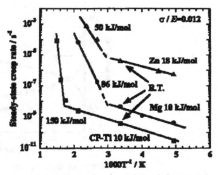

Figure 3. Arrhenius plot of CP-Ti, Mg, and Zn.

229

$$\varepsilon = \varepsilon_i + \varepsilon_p \ln(1+\beta_p t)+\varepsilon_s t \qquad (1)$$

In that equation, ε is the true strain, ε_i is the instantaneous strain, ε_p and β_p are parameters characterizing the primary creep region, ε_s is the extrapolated steady state creep rate, and t is the elapsed time.

Figure 3 shows an Arrhenius plot of the obtained steady state creep rate of the three materials. This plot shows two regions. One is the low temperature dislocation creep region, which has larger apparent activation energies of 50–150 kJ/mol. Another is an ambient temperature creep region that shows very low apparent activation energies of 10–18 kJ/mol. For the three materials, room temperature belongs to the ambient temperature region, even though room temperature is 0.43 of the melting temperature for Zn.

For this study, TEM observations were performed after creep time of about 45 000 s. At that period, the primary creep is ending and the secondary creep is starting, as shown in Fig. 4, which depicts the extrapolated steady state creep rate as a dotted line arrow.

Associated TEM bright-field images are shown in Fig. 5 showing a similar dislocation structure: straightly aligned dislocation arrays without any tangled dislocation. The Burgers' vectors of the three samples were identified as being the same direction, $<\bar{1}2\bar{1}0>$. Then, dislocation line directions were $<0001>$ for CP-Ti and Mg and $<\bar{1}2\bar{1}0>$ for Zn. Therefore, the activated dislocations and slip systems were edge dislocations and prismatic slip for CP-Ti, edge dislocations and prismatic slip for Mg and screw dislocations and basal slip for Zn, as summarized in Table II.

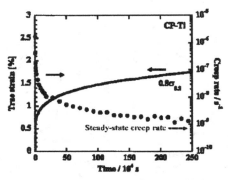

Figure 4. A creep curve and creep rate of CP-Ti. The dotted line arrow
is the steady-state creep rate extrapolated by equation (1).

230

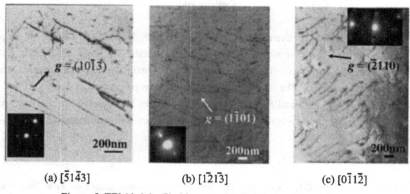

(a) [$\bar{5}1\bar{4}3$] (b) [$1\bar{2}1\bar{3}$] (c) [$0\bar{1}1\bar{2}$]

Figure 5. TEM bright-filed images: (a) CP-Ti, (b) Mg, and (c) Zn

Discussion

Ambient temperature creep was observed in all three metals and was not related with c/a ratios. The apparent activation energies were very low, 10–18 kJ/mol, which are much lower than those of dislocation core diffusion and volume diffusion. Therefore, we consider that unknown accommodation mechanisms are operating during the creep.

Dislocations and slip systems activated in ambient temperature creep for CP-Ti and Zn resembled those in conventional room temperature deformation. For that reason, loads during ambient temperature creep were not sufficiently high to activate other slip systems. However, for Mg, secondary slip was activated in ambient temperature creep. Actually, Mg has an ideal c/a ratio. Therefore, the activated slip system might be determined using the Shumid factor. A deformation texture (0001)-[$1\bar{2}1\bar{0}$] was introduced easily by hot rolling in Mg. In this paper, samples were machined parallel to the rolling direction. Therefore, a primary slip system was not activated.

The TEM observations revealed straightly aligned dislocation arrays and no tangles. One slip system was activated during ambient temperature creep. Cross slippage was not apparent in the HCP structure. Therefore, the work-hardening speed was slow and creep deformation continued.

Table II. Slip Systems Activated During Ambient Temperature Creep

Sample	Dislocation			Slip System
	Type	b	Line Direction	
CP-Ti	Edge Dislocation	$<\bar{1}2\bar{1}0>$	$<0001>$	Prismatic Slip
Mg	Edge Dislocation	$<\bar{1}2\bar{1}0>$	$<0001>$	Prismatic Slip
Zn	Screw Dislocation	$<\bar{1}2\bar{1}0>$	$<\bar{1}2\bar{1}0>$	Basal Slip

231

Conclusions

In this study, we identified active dislocations in the representative HCP metals, i.e., CP-Ti, pure Mg and Zn, using TEM after ambient temperature creep tests. The TEM observation results showed that all samples had similar dislocation structures: straightly aligned dislocation arrays, but no dislocation tangles. The identified slip systems were prismatic slip for CP-Ti and the prismatic slip for Mg and basal slip for Zn. One slip system was observed in a crystal grain, indicating that cross slip and work hardening do not occur; then deformation proceeded.

References

1. A.W. Thompson and B.C. Odegard, "The Influence of Microstructure on Low Temperature Creep of Ti-5Al-2.5Sn," *Metall. Trans. A*, 4 (1973), 899-908.

2. B.C. Odegard and A.W. Thompson, "Low Temperature Creep of Ti-6Al-4V," *Metall. Trans. A*, 5 (1974), 1207-1213.

3. S. Ankem, C.A. Greene, and S. Singh, "Time Dependent Twinning During Ambient Temperature Creep of A Ti-Mn Alloy," *Scripta Metall.*, 30 (6) (1994), 803-808.

4. T. Neeraj, D.H. Hou, G.S. Daehn, and M.J. Mills, "Phenomenological and Microstructural Analysis of Room Temperature Creep in Titanium Alloys," *Acta Metall.*, 48 (2000), 1225-1238.

5. A.W. Cottrell, "Criticality in anrade creep," *Philos. Mag.*, 74 (1996), 1041-1046.

6. E. Sato, T. Yamada, H. Tanaka, and K. Kuribayashi, "Categorization of Ambient-temperature Creep Behavior of Metals and Alloys on their Crystallographic Structure," *Mater. Trans.*, 47 (4) (2006), 1121-1126.

Innovations in
Titanium
Technology

Microstructure and Properties II

High Temperature Oxidation of Ti₃Al-4at%Nb Alloy in Pure Oxygen

Chris Williams[1], D. Mantha[2] and R. G. Reddy[3]

[1]*Graduate Student,* [2]*Research Engineer and* [3]*ACIPCO Professor and Head*
Department of Metallurgical and Materials Engineering
The University of Alabama, Tuscaloosa, AL 35487-0202

Keywords: Titanium aluminum intermetallics, oxidation, Ti₃Al

Abstract

Oxidation of Ti₃Al-4.0at%Nb alloy in pure oxygen was investigated. The samples were oxidized in the temperature range 1023-1373 K using thermogravimetric analyzer (TGA) for 24 and 48 hours. Effective activation energy (Q) for the temperature range 1023-1173 K was 370.5 kJ/mole and for the temperature range 1173-1373 K was 97.9 kJ/mole, as determined from the weight gain data. This composition showed a distinct change in oxidation mechanism at 1173 K, which needs further study. Oxidation rate increased as the temperature was increased. Reaction products formed are rutile and alumina.

Introduction

Titanium-aluminum intermetallics have received much attention due to their excellent mechanical properties. The combination of high strength, low density, and good creep properties is of particular interest. Unfortunately, these alloys suffer from oxidation at high temperatures. Various alloying elements have been tested in an attempt to improve the oxidation resistance of these alloys in various atmospheres [1-8]. This improvement in oxidation properties would be achieved when the oxidation products are thermodynamically stable, slow growing, and adherent to the alloy [6].

Experimental Procedure

Materials and Sample Preparation

Titanium rod of >99.99% purity, aluminum ingot of >99.99% purity, and niobium rod of >99.99% purity were used in preparing the Ti₃Al-4.0at%Nb alloy. Quantities were cut from each such that the resulting composition would fall into the Ti₃Al phase range. The composition selected was Ti-25%Al-4.0%Nb. The cut pieces were placed in an arc-melter and melted three times for homogeneity. The resulting ingot was allowed to cool to room temperature before it was removed from the arc-melter, and impurities from the crucible were ground off. Ingot was wrapped in tantalum foil and encased in a quartz tube under vacuum and annealed for one week at 1273 K. Annealed ingot was cut into 5 x 5 x 2 mm samples by EDM wire cutting machine. Samples were polished with 400 and 600 grit SiC papers followed by polishing with diamond slurry on cotton cloth. Polished samples were ultrasonically cleaned in acetone bath to remove any polishing media or other impurities.

Oxidation Testing

Oxidation experiments were conducted using Perkin Elmer TGA 7 HT hardware with Pyris software. Ti_3Al-4.0at%Nb samples were placed in a platinum sample pan. Isothermal experiments were conducted in the temperature range 1023-1373 K at an interval of 50 K for 24 and 48 hours. High purity argon was passed through the reaction chamber while heating to experimental temperature and also while cooling to room temperature to prevent oxidation during non-isothermal region. High purity argon purge gas at a flow rate of approximately 3 cm^3 per second was switched to ultra high purity oxygen (99.999 % pure) at a flow rate of approximately 3 cm^3 per second for the duration of the isothermal period. Time, sample temperature, and sample weight gain was measured once per minute by the Pyris software during the experiment. Samples were carefully removed from the platinum pans after the experiments and labeled for characterization. Oxidized samples were characterized by Philips X-ray Diffractometer using Cu target with K_α as wavelength and scanning speed of $2\theta = 0.02°$ per second and Philips XL30 Scanning Electron Microscope.

Results and Discussion

Oxidation experiments of Ti_3Al-4.0at%Nb samples were conducted in the temperature range of 1023-1373 K in pure oxygen. Samples showed greater oxidation rates with increase in temperature, as seen in figures 1 and 2. Figure 1 shows the weight gain per unit area versus time for the samples run at temperatures from 1023-1223 K. Figure 2 shows the weight gain per unit area versus time for the sample run at 1373 K.

The data in figure 2 was plotted separately from that in figure 1 due to the difference in scale. The weight gain per unit surface area was more than an order of magnitude larger in the 1373 K sample than the other samples. This allows comparison of the data in figure 1. This increase in the oxidation kinetics compared to the other samples is thought to be due to a combination of the increase in oxygen diffusion with increased temperature as well as the predominately open rutile microstructure of the 1373 K sample allowing for easier diffusion paths than the more compact mixed alumina and rutile oxide scales formed at lower temperatures [9].

Another method for comparing the oxidation rates is the "parabolic rate constant," (k_p) [10]. This value was obtained by plotting time versus weight gain per unit surface area, then fitting a second order equation to this for each temperature. This yields an expression in the form of equation 1:

$$t = A + B\ (\Delta W)/S + C\ ((\Delta W)/S)^2 \qquad (1)$$

Where t is time, $(\Delta W)/S$ is the weight gain per unit surface area, and A, B and C are constants. Reciprocal of C gives k_p. This determination is irrespective of the oxidation mechanism. The resulting k_p values are plotted in the form of $\ln(k_p)$ as a function of reciprocal temperature $(1/T)$. This plot is shown in Figure 3.

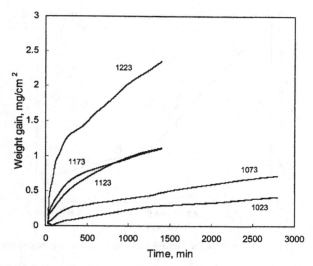

Fig. 1: Weight gain per unit surface areas versus time for Ti3Al-4.0at%Nb samples run from 1023-1223 K for 24 and 48 hours.

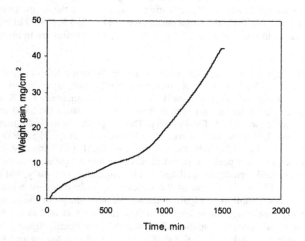

Fig. 2: Weight gain per unit surface area versus time for Ti3Al-4.0at%Nb sample run at 1373 K for 24 hours.

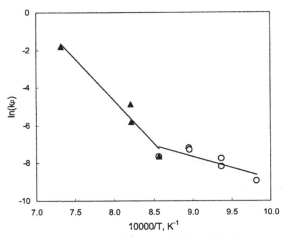

Fig. 3: Ln (kp) vs. 10000/T for Ti₃Al-4.0at%Nb oxidized in pure oxygen.

Researchers have calculated the activation energies of the alloys using a function of the slope of a single linear fit to the k_p data [1, 2, 4-7, 11, 12]. Ti₃Al-4.0at%Nb appears to poses a break in this linear relationship at 1173 K; Two activation energies were determined for the temperature ranges 1023-1173 K and 1173-1373K: activation energy, Q, for the temperature range 1023-1173 K is 370.5 kJ/mol and Q for the temperature range 1173-1373 K is 97.9 kJ/mol. This can be explained by a change in the mechanism for oxidation. Further studies are in progress to explain this mechanism.

The SEM and XRD data show that Al_2O_3 predominates in the range 1073-1173 K, whereas TiO_2 predominates above 1173 K. At 1023 K, relatively small oxide layer is formed, which is a mixture of rutile and alumina. Different oxidation products formed at 1173 K also suggests a change in oxidation mechanism at this temperature. Figure 4 shows the SEM image of Ti₃Al-4.0at%Nb alloy oxidized at 1023 K for 48 hours. The image shows rutile (lighter particles) and alumina (darker particles) grow on the base Ti₃Al-4.0at%Nb alloy. The XRD pattern of the sample was presented in figure 5. High intensity peaks of Ti₃Al, TiO_2, and Al_2O_3 were observed at 1023 K. However, niobium peaks were not observed at any temperature in the current study. The oxide scale exhibited significant spallation at all temperatures of study, but was reduced at temperatures above 1173 K. Spallation of the scales was easily observed while removing the samples from the TGA sample pan. Figure 6 shows the SEM image of the sample oxidized at 1073 K for 48 hours. The image shows a crack in the oxide scale as well as the re-growth of the oxide scale where it had previously spalled off during the experiment. Figure 7 shows the XRD pattern of the sample oxidized at 1073 K for 48 hours. Alumina is the major oxide, with some rutile. For all temperatures, the XRD pattern may show a higher composition of Ti₃Al than is present in the oxide scale due to spallation of the oxide scale.

Fig. 4: SEM image for Ti₃Al-4.0at%Nb sample oxidized at 1023 K for 48 hours.

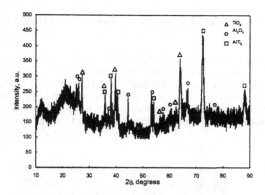

Fig. 5: XRD pattern for Ti₃Al-4.0at%Nb sample oxidized at 1023 K for 48 hours.

Fig. 6: SEM image for Ti₃Al-4.0at%Nb sample oxidized at 1073 K for 48 hours.

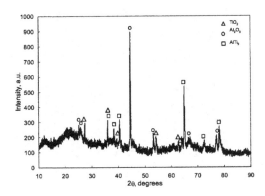

Fig. 7: XRD pattern for Ti₃Al-4.0at%Nb sample oxidized at 1073 K for 48 hours.

Figure 8 shows the SEM image for Ti₃Al-4.0at%Nb sample oxidized at 1123 K for 24 hours. It shows cracking in the oxide scale, as well as re-growth of the oxide scale in a region that had spalled off. Figure 9 shows the XRD pattern of the sample oxidized at 1123 K for 24 hours. It reveals Ti₃Al and Al₂O₃ to be the majority of the sample surface, with some TiO₂. Figure 10 shows the SEM image of the sample oxidized at 1173 K for 24 hours. It shows the Al₂O₃ scale on the left, and a newly spalled region on the right. Figure 11 shows the XRD pattern for Ti₃Al-4.0at%Nb sample oxidized at 1173 K for 24 hours. This sample also shows a very strong Al₂O₃ presence, followed by Ti₃Al and TiO₂.

Fig. 8: SEM image for Ti₃Al-4.0at%Nb sample oxidized at 1123 K for 24 hours.

240

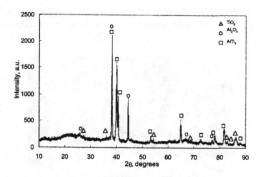

Fig. 9: XRD pattern for Ti₃Al-4.0at%Nb sample oxidized at 1123 K for 24 hours.

Fig. 10: SEM image for Ti₃Al-4.0at%Nb sample oxidized at 1173 K for 24 hours.

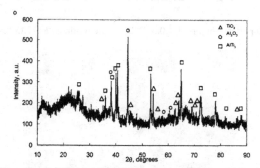

Fig. 11: XRD pattern for Ti₃Al-4.0at%Nb sample oxidized at 1173 K for 24 hours.

Figure 12 shows the SEM image for Ti₃Al-4.0at%Nb sample oxidized at 1223 K for 24 hours. The image shows large, plate-like rutile structures interspersed on a layer of alumina. It is at this point that rutile growth begins to dominate. Figure 13 shows the XRD pattern of the sample oxidized at 1223 K for 24 hours. TiO_2 is the main oxide formed, but Al_2O_3 is more prevalent than the base metal. Figure 14 shows the SEM image for Ti₃Al-4.0at%Nb sample oxidized at 1373 K for 24 hours. This shows rutile crystals that have grown to cover the entire surface, and

241

little or no alumina is seen. Figure 15 shows the XRD pattern of the sample oxidized at 1373 K for 24 hours. TiO_2 makes up the largest portion of the oxide scale, but Al_2O_3 is still present.

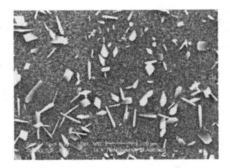

Fig. 12: SEM image for Ti_3Al-4.0at%Nb sample oxidized at 1223 K for 24 hours.

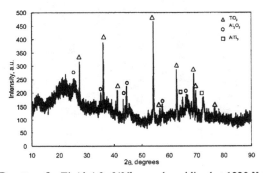

Fig. 13: XRD pattern for Ti_3Al-4.0at%Nb sample oxidized at 1223 K for 24 hours.

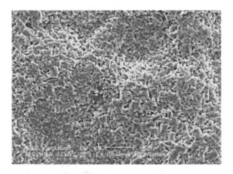

Fig. 14: SEM image for Ti_3Al-4.0at%Nb sample oxidized at 1373 K for 24 hours.

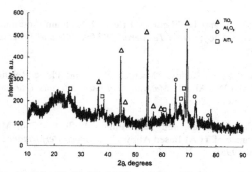

Fig. 15: XRD pattern for Ti₃Al-4.0at%Nb sample oxidized at 1373 K for 24 hours.

Conclusions

Oxidation rate of Ti₃Al-4.0at%Nb alloy increased with increase in temperature. No dense, adherent, and protective alumina layer was formed. The oxide layer spalled off frequently to reveal the Ti₃Al alloy; the strong Ti₃Al peaks in many XRD patterns may be due to this Ti₃Al being revealed rather than representing the composition of the oxide scale. The oxidation mechanism changes above the oxidation temperature of 1173 K, with alumina being the predominant oxidation product formed in the range 1023-1173 K, while rutile being the predominant oxidation product above 1173 K. Effective activation energy (Q) for the temperature range 1023-1173 K was 370.5 kJ/mol and for the temperature range 1173-1373 K was 97.9 kJ/mol.

Acknowledgements

The authors are pleased to acknowledge the financial support for the research by National Science Foundation, Grant No. DMR-0312172 and The University of Alabama.

References

1. Livingston, D., D. Mantha, and R. G. Reddy, "Oxidation Kinetics of Ti₃Al-2.63 at % Nb Alloy," *High Temperature Materials and Processes.* vol. 24, no. 5 (2005) pp. 259-267.

2. Reddy, Ramana G., Yang Li, and Mario F. Arenas, "Oxidation of a Ternary Ti₃Al-Ta Alloy," *High Temperature Materials and Processes.* vol. 21, no. 4 (2002) pp. 195-205.

3. Anada, Hiroyuki, and Yoshiaki Shida, "Effect of Mo Addition on the Oxidation Behavior of TiAl Intermetallic Compound," *Materials Transactions, JIM.* vol. 36, no. 4 (1995): pp. 533-539.

4. Reddy, Ramana G., "In-Situ Multi-layer Formation in the Oxidation of Ti₃Al-Nb," *JOM*, (February 2002): pp.65-67.

5. Reddy, R. G. and Y. Li, "Isothermal Oxidation of a TiAl-Nb Alloy," *High Temperature Materials and Processes.* vol. 20, nos. 5-6 (2001): pp. 319-331.

6. Reddy, R. G., Y. Li, and D. Mantha, "Effect of Niobium Addition of the Oxidation Behavior of a Ti₃Al Alloy," *High Temperature Materials and Processes*. vol. 22, no. 2 (2003): pp. 73-85.

7. S. G. Kumar and R. G. Reddy, "Microstructure and Phase Relations in a Powder-Processed Ti-22Al-12Nb Alloy," *Metallurgical and Materials Transactions A*. vol. 27A (April 1996): pp. 1121-1126.

8. Yufeng, Sun, Cao Chunxiao, and Yan Mingao, "Effect of Nb, Mo, V Contents on Oxidation Resistance of Ti₃Al Based Alloys," *Advanced Performance Materials*. vol. 2, no. 3 (September 1995): pp. 281-288.

9. Reddy, R.G., X. Wen, and I.C.I. Okafor, "Interdiffusion in the TiO2 Oxidation Product of Ti₃Al," *Metallurgical and Materials Transactions*. vol. 32A (March 2001): pp. 491-495.

10. D. Monceau and B. Pieraggi: *Oxidation of Metals*, Vol 50, Nos 5/6. 477 (1998).

11. Mantha, D., X. Wen, and R. G. Reddy, "High Temperature Oxidation of a Ti₃Al Alloy in Argon-5% SO₂ Environment," *High Temperature Materials and Processes*. vol. 23, no. 2 (2004): pp. 93-101.

12. Reddy, R.G., X. Wen, and M. Divakar, "Isothermal Oxidation of TiAl Alloy," *Metallurgical and Materials Transactions A*. vol. 32A (September 2001): pp. 2357-2361.

EVALUATION OF TIN COATINGS PRODUCED VIA DIFFERENT TECHNIQUES

Ali Arslan Kaya, Selda Ucuncuoglu and Kerim Allahverdi

TUBITAK MRC, Materials Institute
P.O.B. 21, Gebze 41470, Kocaeli, TURKEY

Keywords: Titanium Nitride, Hard-alpha, Confocal Raman Spectroscopy

Abstract

TiN coatings were created on Ti-6A-l4V substrate using two different techniques, namely PVD and heat treatment under controlled atmosphere. The aim was to choose one of these techniques to employ for creating TiN deposits of known size, shape and location on the surface of a titanium alloy block. This surface was later to be bonded to another titanium alloy block via diffusion bonding, thus encapsulating the TiN deposits in the metallic body. Such titanium bodies carrying TiN as synthetic defects can be used as calibration blocks in ultrasonic inspection of titanium alloy parts in aircraft industry. The TiN coatings produced via two methods were characterized using Confocal Raman, Atomic Force Microscopy (AFM), XRD, scratch test as well as scanning electron microscopy (SEM) techniques.

Introduction

Titanium alloys serve in a wide range of applications because of their congruency for high stress conditions and oxidation resistance. However, these materials show weak wear resistance and high friction coefficient, demanding surface improvements to be coupled with their desirable mechanical properties. There exist a large number of processes to induce surface modifications on materials. Physical Vapour Deposition (PVD) and transformation of surfaces via heat treatment are two of the available methods [1]. Since TiN coating via heat treatment is a result of a conversion of the material surface we will be using the term 'conversion coating' throughout the rest of this report.

TiN coatings are considered as high technology application and commonly used in micro-electrics, aerospace technology, and semiconductor industry due to their unique characteristics such as high hardness, improved friction performance, excellent wear, corrosion and erosion resistance, good thermal conductivity, higher electric conductivity, higher melting temperatures and diffusion barrier properties [2-4]. However, essentially the same phase is totally undesirable due to its embrittling effect in cast and/or heat-treated load-bearing components in critical applications such as aircraft engine parts. The aim of this study is an evaluation of TiN coatings formed via PVD and heat treatment techniques. PVD TiN coatings are commercially widely used in industrial applications, while conversion coating is not. The idea of comparing the two different methods in terms of the characteristics of TiN stemmed from the need to choose a method that would give a representative TiN structure comparable to TiN type defects that could form in actual cast and/or heat-treated structures of titanium alloys. The chosen method would then be used to form TiN in desired size, shape and location on a titanium alloy block to be encapsulated within. Such an alloy block would serve the purpose of a calibration block in ultrasonic inspection of titanium alloy parts provided that the encapsulated TiN properties are maintained at the end of the whole manufacturing processes involved. Concerns regarding the characterization of TiN formed via different processes can thus be justified. Evaluations on TiN structures in this study was done by employing various techniques, namely Confocal-Raman microscopy, Atomic Force Microscopy (AFM), scratch test, XRD and SEM.

Materials and Methods

The chemical composition of the Ti-6Al-4V alloy is given in Table 1. Ti6Al4V alloy surface was polished to a mirror finish prior to coating deposition. Deposition of TiN coating (~1.5μm thick) was done using cathodic arc-evaporated PVD at a commercial facility. In arc vapor deposition the vapor to be deposited is formed from an electrode bearing a low-voltage high-current DC arc in a low-pressure gaseous atmosphere. In cathodic arc vaporization, which is the most common PVD arc vaporization process, the high current density arc moves over a solid cathodic electrode causing local heating and vaporization. High ionizing rate and high bombardment energy are unique technical advantages of cathodic arc deposition. [5,6]. The conversion coating TiN (120μm thick) was formed via heat treatment under nitrogen atmosphere (formed via gas flow) at 1250°C for 8h followed by air-cooling. The cross sections of TiN coated Ti-6Al-4V samples for examination were prepared via standart metallographic tecniques, i.e.sectioning and cold mounting in epoxy followed by grinding with SiC paper, polishing with 1micron diamond paste and finally with 0.5 micron coolidal silica.

Table 1. The chemical composition of the Ti-6Al-4V alloy used in the present investigation (wt%).

Element	Al	V	C	Fe	N	O	H	Ti
Wt%	5.5 – 6.5	3.5 – 4.0	0.08 max	0.25 max	0.05 max	0.13 max	0.012 max	balance

The techniques used to examine the PVD and conversion coating TiNs were Confocal Raman Spectroscopy, Atomic Force Microscopy (AFM), X-ray Diffraction, scratch test and Scanning Electron Microscopy (SEM). Scratch test were performed using commercially available testing equipment (model: CSM microscratch tester) with an acoustic detector and the experimental conditions are the diamond tip radius of 200μm, the scratching speed of 0.25 mm/min, the loading rate of 3N/min and the end load of 30N. AFM was used to investigate the surface roughness of each type of TiN coatings. The AFM scanning was performed on scan are of 40x40 μm for each sample with scan rate of 3 Hz respectively (model:Quesant USPM). SEM studies were conducted by using a JEOL FEG SEM 6335 equipped with an OXFORD Inca EDS microanalytical unit. XRD studies were undertaken with Shidmatzu XRD 6000 model equipment.

Results and Discussions

Raman spectroscopy gives vibrational information about the present bond structure in organic and inorganic molecules. In this method, a laser photon is scattered by a molecule and loses or gains energy during the process. The amount of energy lost is detected as a change in energy of the irradiating photon. This energy loss is characteristic for each particular bond in the molecule and thus serves as a precise spectral fingerprint, unique to a molecule being examined. Raman spectroscopy can also be used to control the uniformity of the thickness of a film. [7,8]. Raman spectrums obtained from PVD and conversion coating TiN structures are given in Fig.1 in which common lines in both spectra are marked. As seen in this figure there exist extra Raman lines in the spectrum of the PVD TiN coating. Further differences can be noticed in some peak heights as well. Such differences that can only be related to the existing

246

bond structure in the two TiN coatings may be attributed to the presence of some oxygen in the conversion coating type.

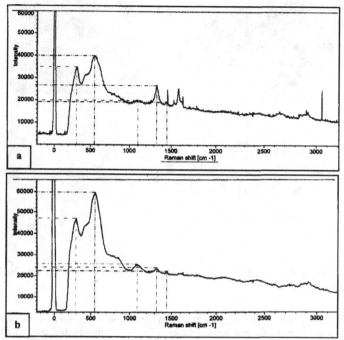

Figure 1. Confocal Raman spectra from (a) PVD TiN coating (b) conversion coating TiN.

Figure 2. shows the three-dimensional surface morphologies of TiN coatings produced by different techniques with 40X40 μm scan areas by AFM analysis. As may be seen from these figures there exist no considerable difference between the two types of coatings in terms of roughness. As the precursor surfaces for both types of coatings were prepared to the same surface finish via identical procedures lack of a noticeable difference between the two after TiN layers were formed is not unexpected.

Scratch adhesion testing is a commonly used technique for determination of mechanical properties of thin surface coatings [9]. This test method utilize a single-point contact, that is a Rockwell C 120 diamond cone, and can induce plastic deformation both in the coating and substrate. In the scratch test a stylus is drawn over the sample surface under a stepwise or continuously increasing normal force until the coating is detached. In this test method the coating detachment can be observed in practice by optical microscopy or scanning electron microscopy [10].

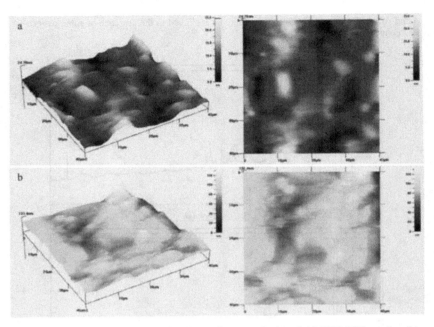

Figure 2. Three-dimensional AFM and top view topography of (a) PVD TiN coating (b) conversion coating TiN.

The scratch responses of the coatings have been illustrated in Fig.3 and 4. Neither of the coatings responded in a fashion that may be described by plastic deformation with pile-up of substrate material around the scratch. Instead, especially PVD TiN coating showed a more smooth and continuous scratch mark whereas the conversion coating a more discontinuous appearance. The friction coefficient and frictional force profiles in Fig. 4 also relates to the scratch images in terms of smoothness and discontinuity for the corresponding coating types. This result may be interpreted as a more brittle nature possessed by the conversion coating that resulted in micro chippings and lack of continuity along the scratch profiles. As seen in the penetration depth profiles the penetration of the tip occurred to a lesser extend in conversion coating compared to PVD type. The detailed morphology of the scratch channels as revealed by scanning electron microscopy is given in Fig.5. Considering the SEM images given in Fig. 5 and 6 the appearance of the scratch marks may be better understood. SEM micrographs in Fig.6 shows that the surface of the conversion coating has a micro morphology related to the grain structure of the precursor alpha phase that was converted to TiN while PVD coating displays a featureless surface. Thus the discontinuity scratch mark may be attributed to the skipping of the scratch tip over the micro features of the conversion coating. The cross-sectional SEM images in Fig. 7 reveals the thicknesses of the coatings as ~1.5μm and 120μm for PVD and conversion coatings, respectively.

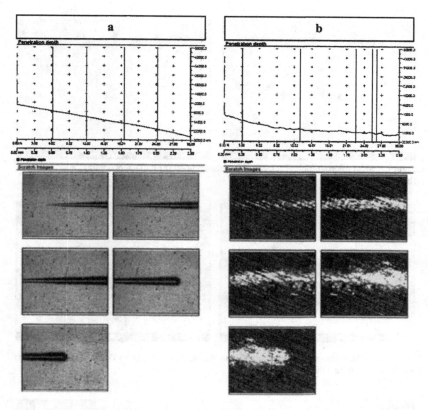

Figure 3. Penetration depth profiles and scratch images of (a) PVD TiN coating; and (b) conversion coating TiN.

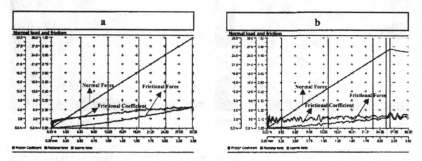

Figure 4. Normal load and friction signal variations during the scratch test (loading rate: 3N/min, scratching speed: 0.25 mm/min, end load: 30N) of (a) PVD TiN coating; and (b) conversion coating TiN.

Figure 5. SEM micrographs of scratch channels made on (a) PVD TiN coating; and (b) conversion coating TiN.

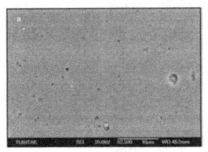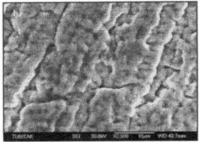

Figure 6. SEM micrographs perpendicular to the surfaces of (a) PVD TiN coating; and (b) conversion coating TiN.

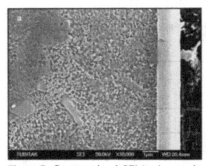

Figure 7. Cross-sectional SEM micrographs of (a) PVD TiN coating; and (b) conversion coating TiN.

X-ray diffraction spectra of PVD and conversion coating TiNs have been shown in Figure 8. As can be seen the major TiN peaks matches for both cases with minor shifts in some positions. Additional peaks belonging to AlTi3 intermetallic in the case of conversition coating were also revealed. This may be a factor increasing the hardness of the conversation coating.

250

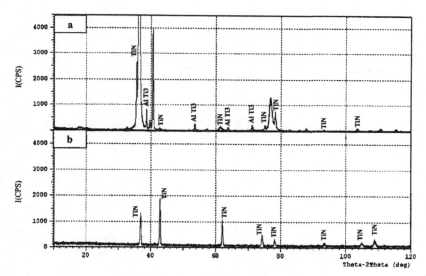

Figure 8. X-ray diffraction spectra of (a) conversion coating, and (b) PVD TiNs.

References

1. G.B. de Souza et al. / Surface & Coatings Technology 191 (2005) 76–82.

2. H.Z. Wu et al. / Thin Solid Films, 191 (1990) 55–67.

3. S.V. Hainsworth, W.C. Soh / Surface & Coatings Technology 163–164 (2003) 515–520.

4. Te-Hua Fang, Sheng-Rui Jlan, Der-San Chuu / Applied Surface Science 228 (2004) 365-
5. 372.

6. Li Zhengyang et al. / Surface & Coatings Technology 131 (2000) 158-161.

7. C.P. Coustable, J.Yarwood, W-D. Munz / Surface & Coatings Technology 116–119 (1999) 155–159.

8. Ingrid De Wolf, Chen Jian, W. Merlijn van Spengen / Optics and lasers in Engineering 36
9. (2001) 213-223.

10. Binnig et al. / Atomic force microscope. Phys. Rev. Lett. (1986) 56(9), 930-933

11. P.J. Burnett and D.S. Rickerby / Thin Solid Films 154 (1987) 403-416.

12. Juhani Valli and Ulla Makela / Wear 115 (1987) 215-221.

SOME ENGINEERING ASPECTS OF THERMOHYDROGEN PROCESSING OF LARGE COMPLEX TITANIUM CASTINGS.

G. Cao[1, 2], H. Nan[2], C. Xie[2]

1 University of Wisconsin-Madison, Madison, WI 53706, USA
2 Institute of Aeronautical Materials, Beijing, 100095, China

Keywords: Thermohydrogen Processing, Titanium, Casting

Abstract

Thermohydrogen processing is an effective method to improve the mechanical properties of titanium alloy castings. In this paper, some engineering aspects of thermohydrogen processing of large complex Ti alloy castings were discussed. From the point of engineering application, the effects of thermohydrogen processing were studied on the overall mechanical properties including room temperature tensile strength, 350°C tensile strength, 350°C durability, room temperature fatigue strength and fracture toughness. The feasibility of applying thermohydrogen processing to large thick-walled (up to 45mm wall thickness) titanium casting was also studied. Some real Ti alloy castings were tested using thermohydrogen processing. Our results showed that thermohydrogen processing was very suitable to heat treat the large complex titanium castings when high temperature durability and fatigue strength are desired or required

Introduction

Since 1980s when Kerr et al [1] found that hydrogen could act as the temporary alloying element in titanium, there have been a lot of reports [2-10] on the effects of hydrogen on the microstructure and mechanical properties of titanium alloys. Several thermohydrogen processing methods were developed. It is agreed that thermohydrogen processing is very suitable to modify the as-cast microstructure and improve the tensile and fatigue strength of titanium alloy castings because they cannot be strengthened by conventional thermomechanical processes. But there are few reports on the effects of thermohydrogen processing on the damage tolerance properties (mainly fracture toughness), high temperature tensile properties, and high temperature durability of titanium alloy castings. The possibility of applying thermohydrogen processing to thick wall large titanium castings was also seldom reported. And there is no public report about the engineering applications of thermohydrogen processing of titanium castings. In this paper, in view of the possible engineering application of thermohydrogen processing (THP), an extensive investigation was carried out on the effects of thermohydrogen processing (THP) on the overall mechanical properties of Ti-6Al-4V alloy, including tensile, fatigue strength, durability and fracture toughness. The possibility of applying THP to thick wall large titanium castings was also studied.

THP process is based on the effects of hydrogen as the temporary alloying element on the phase compositions, development of metastable phases, and kinetics of phase transformation and other chemical reactions. Typical THP process consists of the following steps: (1) hydrogenation, (2) beta solution treatment, (3) eutectic reaction and (4) vacuum dehydrogenation. Because hydrogen is a strong beta stabilizer, the temperature of $\alpha+\beta/\beta$ transformation of Ti-6Al-4V alloy is reduced from about 990°C to 800°C, and the critical cooling rate required for martensite transformation is significantly reduced, which improves the hardenability of Ti-6Al-4V and avoids the probable dimension distortion of castings during THP treatment.

Experimental

Ti-6Al-4V samples were cast via lost wax investment casting. Samples for testing tensile strength, fracture toughness, fatigue strength and high temperature durability were cast. For testing the tensile strength, most samples were machined from standard samples with a cross section of 12mm by 12mm. Some samples were machined from the sprue area with a diameter of 45mm. The Ti-6Al-4V castings for thermohydrogen processing were HIP'ed (hot isostatic pressing) to remove the porosities in the standard samples and internal pipes in samples cut from the sprue areas. After hot isostatic pressing, the samples were sand blasted and or chemically cleaned using HF acid. The equipment for thermohydrogen processing of titanium castings was shown in Fig.1. It mainly contains a seamless stainless steel tube, a movable furnace, and vacuum system (mechanical pump and oil diffusion pump). The Ti-6Al-4V samples and real Ti castings were put at the end of the stainless steel tube and were covered by machined titanium turnings. During heating, the furnace was moved to the position that the stainless steel tube was inserted into the furnace. During cooling, the furnace was removed manually. The hydrogenation temperature is about 700°C and the hydrogenation time depends on the number and thickness of samples. Generally, the hydrogenation time is 1 to 2 hours. After hydrogenation, the temperature is increased to about 860°C and kept at that temperature for 30 minutes. Then the furnace was removed and the stainless steel chamber was cooled in air or using a fan and it was pumped via a mechanical pump to remove the hydrogen gas from the chamber during the cooling process. Then the chamber was heated to 450°C or 550°C and kept at that temperature for about 5 hours during vacuum, and then the temperature was increased to 790°C and a diffusion pump was used to dehydrogenate the samples. The dehydrogenation time is also dependent on the number and thickness of samples, generally, when the degree of vacuum is below 0.001Pa, the hydrogen content in the titanium can be removed below the requirement of specification. A schematic thermohydrogen processing was shown in Fig. 2. During the many steps of the themohydrogen process, we do not need to remove the samples from the seamless stainless steel tube chamber.

After the thermohydrongen processing, the microstructure and mechanical properties were tested.

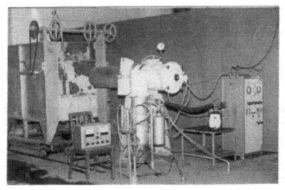

Fig. 1 Thermohydrogen processing equipment

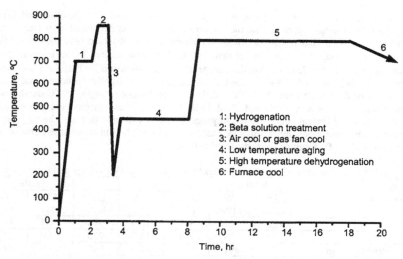

Fig.2 Typical thermohydrogen processing of Ti-6Al-4V castings. Real time for high temperature dehydrogenation will depend on the number and thickness of castings.

Results and discussion

<u>Microstructure and Mechanical Properties of Standard Samples</u>

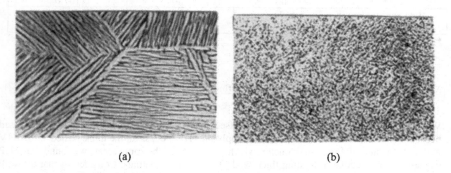

(a)   (b)

Fig.3 Effects of THP on the microstructure of Ti-6Al-4V alloy castings,
(a) as-cast+HIP, (b) after THP

THP can transform the as-cast coarse Widmanstatten microstructure into equiaxed fine-grained morphology and eliminate the continuous grain boundary alpha phase (Fig. 3). Effects of THP on mechanical properties are shown in Table 1 and 2. It can bee seen that the room temperature yield strength of THP'ed samples was increased 14% while the ductility was retained. High temperature tensile properties were improved significantly including UTS and ductility. The high temperature durability and fatigue strength were also increased significantly. The fracture toughness was retained or decreased very slightly and the fracture toughness was still significantly higher than that of beta annealed forgings. Some reports on the effects of THP on the Ti-6Al-4V forgings indicated that the fracture toughness was decreased after treatment. In my research, some similar results on Ti-6Al-4V castings were observed (not shown) when the dehydrogenation temperature was lower than 700°C. But when treated by an innovative THP process as shown in Fig.2, where the dehydrogenation temperature is 790°C, the good fracture toughness of as cast Ti-6Al-4V is maintained or just very slightly reduced. And a good combination of RT tensile strength, 350°C tensile strength, fatigue strength, 350°C durability and fracture toughness could be obtained. And furthermore, all these mechanical properties are superior to those of typical beta annealed forgings. It can be expected that after being processed by thermohydrogen processing, Ti-6Al-4V castings can be applied in those stress critical areas where only Ti-6Al-4V forgings are considered before.

Table 1 Comparison of tensile properties between Ti-6Al-4V castings and forgings

Material	Room temperature tensile				350°C tensile		
Condition	UTS (MPa)	YS (MPa)	EL (%)	RA (%)	UTS (MPa)	EL (%)	RA (%)
HIP	922	861	7.8	15.9	574	11.4	31.5
THP	1015	984	8.0	15.0	720	14.0	54.3
Beta annealed forgings	953	891	14.6	45.3	665	16.0	58.1

Table 2 Comparison of high temperature durability, fatigue strength and fracture toughness of Ti-6Al-4V castings and forgings

Material Condition	Durability 350°C,100h (MPa)	10^7 cycle fatigue* (MPa)	Fracture toughness, K_{IC} (MPa√m)
HIP	539	470	103.6
THP	>700	750	101.0
Beta annealed forgings	657	539	75.9

*: Test condition, R=0.1, Kt=1, f=130Hz

Mechanical Properties of Thick Titanium Castings

When the thickness of the castings is increased, the mechanical properties will be decreased because of the coarse microstructure. After applying THP process on the thick walled Ti-6Al-4V with a dimension of 45mm in diameter, which was cut from the sprue areas, we found that THP was also very effective in treating thick wall titanium alloy castings. Considering that the wall thickness of most large titanium alloy castings was thinner than 45mm, it is highly probable to treat large thick-wall complex titanium alloy castings via THP process. The effects of THP process on the mechanical properties of thick titanium castings were shown in Table 3. From Table 3, we can see the significant improvement in mechanical properties. For the thick Ti-6Al-4V samples, the UTS at room temperature was very low, much lower than that of standard samples with a 12mm by 12mm section size. After THP process, the RT tensile strength was increased a lot. Especially the fatigue strength was increased 70% (from 385MPa for HIP to 650MPa for THP) and even in this case, the tensile strength is comparable to beta annealed forgings and the fatigue strength was higher than that of beta annealed forgings (650MPa vs.

256

539MPa). These results mean that thermohydrogen processing is especially suitable to heat treat thick wall titanium castings.

Table 3 Mechanical properties of Ti-6Al-4V of 45 mm section size

Material condition	Room temperature Tensile				350°C tensile			10^7 fatigue strength (MPa)*
	UTS (MPa)	YS (MPa)	EL (%)	RA (%)	UTS (MPa)	EL (%)	RA (%)	
HIP	857	805	6.4	20.4	519	12.2	39.3	385
THP	956	910	7.6	25.9	641	14.4	63.5	650

*: Test condition, R=0.1, Kt=1, f=130Hz

THP of Real Castings

In the present study, some small aerospace Ti-6Al-4V castings as shown Fig. 4 were used to test the feasibility of THP process. These castings were processed together with samples for testing mechanical properties. After THP process, there are no dimensional distortion and surface oxidation and no microcrack. Because of the capacity of the small stainless tube chamber, real large castings were not tested, but it can be expected from the results from small castings that thermohydrogen processing can be used in thick wall large castings considering the low cooling rate during the beta solution treatment.

Fig. 4 Real titanium castings treated by thermohydrogen processing

Comparison with Other Thermal Treatment Processes for Titanium Alloy Castings

Generally, most titanium alloys castings are applied industrially in as-cast condition or HIPed condition. There were some traditional treatment processes such as ABST , BST or BSTOA. All these heat treating processes can improve the room temperature tensile strength. But because the ABST, BST and BSTOA need very fast cooling during solution process, which will result in possible dimensional distortions especially for large thin walled complex castings. Besides, solution treatment of very thick (eg. 45mm diameter) casting is very difficult. There is no reports about BST or BSTOA that can heat treat casting of 45mm section size. And there are no reports about the effects of ABST, BST or BSTOA on the high temperature tensile and durabilities. Compared to the thermohydrogen processing, the increase in mechanical properties is not as significant as thermohydrogen processing. It should be mentioned that the thermohydrogen processing is very time consuming especially for thick wall large castings. In the hydrogenation process, it took longer for hydrogen to diffuse into the center of the casting so that the hydrogen content is uniform in the castings. In the dehydrogenation process, it will also take a long time to remove the hydrogen from the thick wall titanium casting. Another disadvantage is that hydrogen gas is an explosive gas, one should be very careful in hydrogenation and solution treatment steps during thermohydrogen processing of titanium castings.

Conclusion

THP is an effective process to strengthen the Ti-6Al-4V castings and the probable dimension distortion during treatment can be avoided. The RT tensile, high temperature tensile, fatigue strength, high temperature durability were significantly improved. The good fracture toughness of the as cast condition was maintained or just very slightly decreased. All the mechanical properties except the ductility of THP treated Ti-6Al-4V castings are superior to those of beat annealed forgings counterpart. It is highly probable to treat large complex titanium alloy castings via THP process. Very thick (45mm diameter) titanium castings can also be processed by thermohydrogen processing. In view of no public report about the industrial application of thermohydrogen processing of titanium castings, the results of this paper can provide useful information for commercialization of THP processing of titanium castings.

References

1 W.R. Kerr, "The Effect of Hydrogen as a Temporary Alloying Element on the Microstructure and Tensile Properties of Ti--6Al--4V", Metall. Trans. A. (16A) (6),(1985), pp. 1077-1087.
2 F. H. Froes, O. N. Senkov, and J. I. Qazi, "Hydrogen As a Temporary Alloying Element in Titanium Alloys: Thermohydrogen Processing," International Materials Reviews. (49), (3-4), (2004), pp. 227-245.
3 H. L. Hou, Z.Q. Li and Y. J. Wang, "Technology of Hydrogen Treatment for Titanium Alloy and Its Application Prospect," Chinese Journal of Nonferrous Metals. (13),(3), (2003), pp. 533-549.
4 F. H. Froes, O. N. Senkov and J. I. Qazi, "Beneficial Effects of Hydrogen As a Temporary Alloying Element in Titanium Alloys: An Overview," 11th International Symposium on Processing and Fabrication of Advanced Materials XI; Columbus, OH; USA; 7-10 Oct. 2002. pp. 295-308.
5 A. A. Il'in et. al., "Effect of Thermohydrogen Treatment on the Structure and Properties of Titanium Alloy Castings," Metal Science and Heat Treatment, (44), (5-6), (2002), pp.185-189.
6 D. Eliezer, et. al., "Positive Effects of Hydrogen in Metals," Materials Science and Engineering, A (280) (2000), pp. 220-224.
7 J. I. Qazi, et. al., "Recycling of Titanium and Ti-6Al-4V Turnings Using Thermohydrogen Processing," Light Metals 2000 as held at the 129th TMS Annual Meeting; Nashville, TN; USA; 12-16 Mar. 2000. pp. 885-889.
8 O. N. Senkov, J. J. Jonas and F. H. Froes, "Recent Advances in the Thermohydrogen Processing of Titanium Alloys" JOM (48), (7), (1996), pp. 42-47.
9 A. A. Ilyin, et. al., "Thermohydrogen Treatment--the Base of Hydrogen Technology of Titanium Alloys," Titanium '95. Vol. III; Birmingham; UK; 22-26 Oct. 1995. pp.2462-2469.
10 F. H. Froes and D. Eylon "Thermochemical Processing (TCP) of Titanium Alloys by Temporary Alloying With Hydrogen," Hydrogen Effects on Material Behavior; Moran, Wyoming; USA; 12-15 Sept. 1989. pp. 261-283.

FLAME SPRAY WELDING OF NICRBSI POWDER ALLOY ON TITANIUM ALLOY SUBSTRATE

Xiaojing Xu[1]

[1]Institute of Advanced Forming Technology, Jiangsu University, 301 Xuefu Rd.; Zhenjiang 212013, P.R.China

Keywords: Titanium alloy, NiCrBSi powder alloy, Flame spray-welding, Microstructure, Properties

Abstract

The present paper deals with a study of flame spray welding of NiCrBSi powder alloy on titanium alloy (Ti-6Al-4V) substrate. In order to overcome the harmful effect of titanium oxide to coating, prior to the spraying a surface pretreatment consisting of ion etching and subsequent electroless deposition of zinc was carried out. In order to increase the diffusion of elements, the time for the coating to stay at liquid state was appropriately lengthened. Due to these modifications, a high quality coating with the thickness of about 1 mm was successfully developed. This coating presented many characterization, such as little pore and inclusion, smooth change in elements distribution and microstructures, good metallurgical bonding in interface, and very high wear-resistance as the result of diffusion of Ti element up to the top layer of the coating.

Introduction

Titanium alloy possesses a remarkable combination of characteristics such as high special strength, superior corrosion resistance, etc. However, its wear resistance is not satisfied, which make it unsuitable for uses in some fields.

Flame spray welding (FSW) of Ni based powder is one of the most versatile and cost-effective techniques for production of wear-resistance coating [1, 2 and 3], however, until now, the application of this techniques to titanium alloy remains little reported. The objective of this investigation is to develop FSW of Ni based powder applicable to titanium alloy. The modified FSW is presented, and the microstructure, hardness and wear-resistance were examined.

Experimental

Ti-6Al-4V alloy was used as the substrate material in this study. NiCrBSi pre-alloyed powders with the nominal chemical composition (in wt. %) of 16.4 Cr, 4.1 B, 4.2 Si, 0.84 C, 1.9 Fe, and balance Ni was used as coating materials. The powder has a size in the range of 100~200 mesh. In order to overcome the harmful effect of titanium oxide to coating, a surface pretreatment consisting of ion etching and subsequent electroless deposition of zinc was carried out prior to spraying. The solution of ion etching and electroless deposition of zinc is HCl, HNO_3, HF, and Balance H_2O, and 40 ml/L HF, 40g/L $ZnSO_4 \cdot 6H_2O$ and Balance H_2O, respectively. The duration for ion etching and electroless deposition of zinc are both about 1~2 min. In order to increase diffusion of element, the time for coating to stay at liquid was appropriately lengthened. As comparison, a conventional FSW is also carried out. The main parameters for FSW were listed in Table I.

Table I. The main parameters for FSW

Ion etching	Conducted for modified FSW; Not conducted for conventional FSW
Zinc electroless deposition	Conducted for modified FSW; Not conducted for conventional FSW
Substrate preheat temperature	230~270
Oxygen pressure	140 kPa
Acetylene pressure	100 kPa
Oxygen flow	310~360 L/h
Acetylene flow	430~560 L/h
Spray distance	100~120 mm
Feed rate	1.5 kg/h
Fusion	1025 ~1065
Fusion time	20~80s for modified FSW; 10s for conventional FSW

The hardness distribution along the depth of coating was measured using Vickers microhardness tester with the load of 2.94 N and the time of 15 s. The microstructure and composition of the coatings were analyzed using JXA-840A type scanning transmission electron microscopy (SEM) equipped with AN10000 type energy dispersive X-ray spectroscopy (EDAX). The phase constituents of top surface of coating were analyzed using D/max-2500PC type X-ray diffractometer (XRD).

The wear resistance of coating was evaluated using a pin-on-disk type friction and wear tester. The coating produced by modified FSW and the titanium alloy substrate (as comparison) were used as the pins. The quenched and tempered bearing steel (AISI 52100) with the hardness of HRC 61 was used as the disks. The used traverse speed was 0.63 m/s, 1.26 m/s and 1.88 m/s, respectively. The used load per area was 17.68 MPa. The wear duration was 3600 s. The wear weight loss was measured by an electrical analytical balance with a sensitivity of 10^{-5} g.

Results and discussion

Microstructure

Figure 1 shows the SEM image and EDAX elemental line analysis of coating. Compared with the coating produced by conventional FSW, the coating produced by modified FSW presents obviously fewer pores, smoother gradual variation of element distribution and longer distance of element diffusion. For the coating produced by modified FSW, the element diffusion is considerably enormous, which is up to the top surface of coating due to the longer duration for coating to stay at liquid.

260

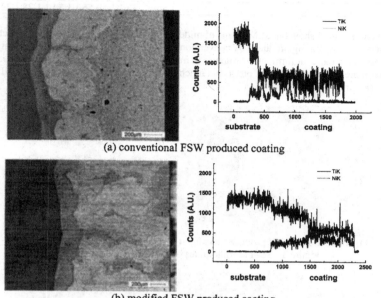

(a) conventional FSW produced coating

(b) modified FSW produced coating

Figure 1. SEM image and EDAX composition line analysis for coating

Figure 2 shows the SEM image of interface zone between coating and substrate. The interface of conventional and modified FSW coating with substrate both presented a metallurgical bonding. The microstructure of interface zone consisted of morphology A and B, which can be determined by EDAX analysis (Table II) to be Ti_2Ni and $Ti_2Ni + (Ti + Ti_2Ni)$, respectively.

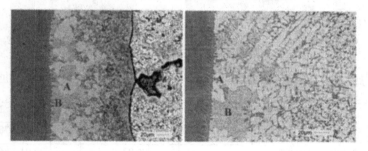

(a) conventional FSW produced coating (b) modified FSW produced coating

Figure 2. SEM morphologies of interface zone

Table II. EDAX analytical result of elemental compositions in interface zone

Specimen	Zone	Compositions (wt. %)						
		C	Al	Si	Ti	V	Cr	Ni
Conventional FSW produced coating	A	–	4.39	1.93	60	1.04	2.9	29.73
	B	–	7.87	3.33	60.61	0.99	2.7	24.5
Modified FSW produced coating	A	30.37	5.01	0.87	41.35	0.3	1.09	21

261

| | B | 14.9 | 5.96 | 3.61 | 50.56 | 1.34 | 1.28 | 22.36 |

Figure 3 and Figure 4 show the SEM image of middle zone and top zone of coating, respectively. Compared with the coating produced by conventional FSW, the coating produced by modified FSW presented a more uniform microstructure in middle zone (Figure 3), a finer microstructure of matrix, more quantity of second phase and much higher content of Ti element (Table III) in top zone (Figure 4).

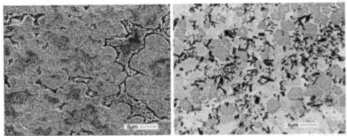

(a) conventional FSW produced coating (b) modified FSW produced coating
Figure 3. SEM morphologies of middle zone of coating

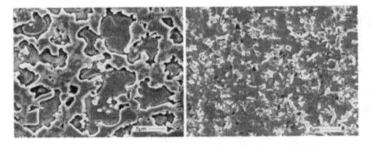

(a) conventional FSW produced coating (b) modified FSW produced coating
Figure 4. SEM morphologies of top zone of coating

Table III. EDAX analytical result of elemental compositions in top zone of coating

Specimen	Compositions (wt. %)						
	C	Al	Si	Ti	V	Cr	Ni
Conventional FSW	–	–	3.55	–	–	25.54	70.91
produced coating	–	–	6.92	–	–	26.91	66.17
Modified FSW produced coating	28.6	0.45	4.35	22.23	3.28	4.84	36.26
	27.01	1.87	3.74	11.46	0.34	5.49	50.1
	30.41	0.8	4.56	5.39	0.75	13.84	44.26

XRD analysis

Figure 5 shows the X-ray diffraction spectrum of top surface of coating. The phase constituents are mainly Ni, $Cr_{23}C_6$ and Ni_3B for the top surface of the coating produced by conventional FSW, whereas the phase constituents are transmitted to be $TiNi_3$, $TiNi$, $Cr_{23}C_6$ and Ni_3B because of higher content of Ti element for the top surface of the coating produced by modified FSW.

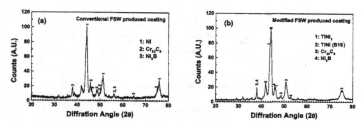

Figure 5. X-Ray diffraction spectrum for the top layer of coatings

Microhardness

Figure 6 shows the variation of microhardness with depth below top surface of coating. Compared with the coating produced by conventional FSW, the coating produced by modified FSW presents a higher hardness and a smoother gradual variation of hardness. The higher hardness can be attributed to the formation of $TiNi_3$ and $TiNi$, which has a higher hardness than Ni. The smoother gradual variation of hardness can be ascribed to the less pores and inclusions.

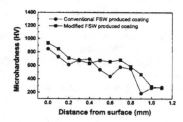

Figure 6. Micro-hardness variation as a function of depth below top surface of coating

Wear properties

Figure 7 gives a comparison of wear mass loss between the modified FSW produced coating and the titanium alloy substrate. It can be seen that the coating improves the wear resistance of the titanium alloy considerably, which is an increase of about three ~ six times.

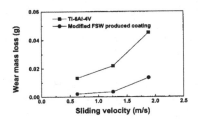

Figure 7. Wear mass loss as a function of sliding velocity

Conclusions

High quality wear-resistance coating on titanium alloy substrate was successfully developed through some modification on conventionally used flame spraying weld technology, using NiCrBSi powder alloy as raw materials. The modified flame spray welding involved in two key steps: one is surface pretreatment prior to spraying, which consists of ion etching and subsequent electroless deposition of zinc; another is to lengthen the duration for coating to stay at liquid during fusing after spraying. As-obtained coating was characterized by few pores and inclusion; a smooth change in microstructure, element distribution and hardness, and a considerable improvement in hardness. This coating presented a about six times improvement in wear resistance compared with substrate titanium at room temperature and under heavy load.

References

1. K. G. Budinski, "Tribological Properties of Titanium Alloys", *Wear,* 151(1991), 203-213.
2. Hyung-Jun Kim,Soon-Yong Hwang,Chang-Hee Lee, and Philipe Juvanon. "Assessment of Wear Performance of Flame Sprayed and Fused Ni-based Coating", *Surface & Coating Technology,* 172(2003), 262-269.
3. N. Kahraman, G. Behcet. "Abrasive Behavior of Powder Flame Sprayed Coatings on Steel Substrates", *Materials & Design,* 23(2002),721-725.

AUTHOR INDEX

W

X

Y

Z

SUBJECT INDEX

T

W